GEOGRAPHY
FOR CCEA GCSE

KAY CLARKE, LYNDA FRANCIS,
PETULA HENDERSON, CORMAC MCKINNEY

Hodder Murray

A MEMBER OF THE HODDER HEADLINE GROUP

ACKNOWLEDGEMENTS

The authors and publishers would like to thank the following individuals, institutions and companies for permission to reproduce copyright illustrations in this book:

Avalon Guitars, page 141. British Wind Energy Association, page 119 (top right and bottom right). Photograph used by permission of Cambridge Science Park, page 137. Colourpoint Books, page 209 (top left). Comic Relief, page 164. The Co-operative Group, page 159. © Corbis, pages 2 (anemometer, barometer, rain gauge, wind vane), 11 (bottom left), 21 (top left), 23 (top left and bottom left), 24 (top right and bottom right), 37 (top right), 40, 55, 60 (bottom left), 62, 63, 74 (left), 77 (left and right), 81 (top), 87 (middle, bottom left and bottom right), 88, 89, 92, 93 (top right and top left), 118 (top), 119 (top left and bottom), 120 (top right and bottom left), 122, 123 (top), 134 (middle), 153, 161, 196 (top right), 197 (left), 198 (top right and top left), 204 (bottom left), 207, 211. The Day Chocolate Company, page 160. © Dinodia, page 154 (bottom right). Meteosat Data © 2003 EUMETSAT, page 15. Miss C Gilmer, Omagh Academy, page 219. N. Rowles and M. Raw, *Total Revision Geography*, 2003, © HarperCollins Publishers Ltd., reprinted by permission of HarperCollins Publishers Ltd., page 27. © The Hindu Photo Archives, page 154 (top left and top right). © Crown Copyright, reprinted by permission from HMSO, page 156. *Earthworks 3*, J. Widdowson, John Murray, 1999, reproduced by permission of Hodder Arnold, pages 222. *GCSE Geography In Focus*, J. Widdowson, John Murray, 2001, Reproduced by permission of Hodder Arnold, page 144. *Cutover and Cutaway Bogs Education Pack*, 2000, by permission from the Irish Peatland Conservation Council (IPCC), Lullymore, Rathangan, Co. Kildare, pages 81 (below), 83. Irwin's Bakery, The Food Park, Carne, Portadown, Co. Armagh, Northern Ireland, pages 125, 128 (left middle and left bottom). Courtesy of Laganside Corporation, page 204 (bottom right). © Crown Copyright Met Office, page 15. © Mirrorpix, page 11 (bottom right). *Geoactive 259*, Nelson Thornes, 2002, page 25 (bottom left). *New Key Geography 1 for GCSE*, by D. Waugh and T. Bushell, Nelson Thornes, 2002, page 157 (top). *Themes in Human Geography* by M. Rocket, Nelson Thornes, 1987, page 157 (bottom left). *The New Wider World*, 2nd Edition by D. Waugh, Nelson Thornes, page 220 (top right). *Exploring Our World: Investigating Issues of Interdependence and Social Justice in the 21st Century*, One World Centre, Belfast, 2001; Statistics taken from the Human Development Report, United Nations Development Programme, New York, 2000, pages 148 (top), 169. Reproduced from the 2003 Ordnance Survey of Northern Ireland 1:50 000 Sheet 5 map with the permission of the Controller of Her Majesty's Stationery Office, © Crown Copyright 2003; Permit number 40206, page 44. Reproduced from the 2002 Ordnance Survey of Northern Ireland 1:50 000 Sheet 9 map with the permission of the Controller of Her Majesty's Stationery Office, © Crown Copyright 2002; Permit number 40206, pages 44, 177. Reproduced from the 2003 Ordnance Survey of Northern Ireland 1:50 000 Sheet 15 map with the permission of the Controller of Her Majesty's Stationery Office, © Crown Copyright 2003; Permit number 40206, page 129. © Oxford University Press 1998 from *Complete Geography* by Simon Chapman *et al.* (OUP, 1998), reprinted by permission of Oxford University Press, pages 147, 148 (bottom). Stephen Roulston © Geography In Action, pages 25, 203 (top left). © Science Photo Library, page 2 (barograph, Stevenson's screen and max-min thermometer). © Waste Watch, page 209 (top left). World Bank (c/o the Copyright Clearance Centre), page 150. Yorkgate Retail & Leisure Park, pages 130, 131, 132.

The authors and publishers would also like to thank the following for permission to reproduce material in this book:

The Co-operative Group for an extract from the Co-op Fair Trade information booklet, used on page 161 (bottom right). *The World: Places and Cases* by R. Prosser, Nelson Thornes, page 144 (bottom right). One World Centre, Belfast for extracts from *Exploring Our World: Investigating Issues of Interdependence and Social Justice in the 21st Century* (2001), used on page 161 (top, and bottom left). *Western Education & Library Board* for extracts from *Thinking through Geography* pages 6, 15, 127.

Every effort has been made to trace and acknowledge ownership of copyright. The publishers will be glad to make suitable arrangements with any copyright holders whom it has not been possible to contact.

Note about the Internet links in the book. The user should be aware that URLs or web addresses change regularly. Every effort has been made to ensure the accuracy of the URLs provided in this book on going to press. It is inevitable, however, that some will change. It is sometimes possible to find a relocated web page, by just typing in the address of the home page for a website in the URL window of your browser.

Orders: please contact Bookpoint Ltd, 130 Milton Park, Abingdon, Oxon OX14 4SB. Telephone: (44) 01235 827720. Fax: (44) 01235 400454. Lines are open from 9.00–6.00, Monday to Saturday, with a 24 hour message answering service. You can also order through our website www.hoddereducation.co.uk.

British Library Cataloguing in Publication Data
A catalogue record for this title is available from the British Library

ISBN-10: 0 340 86918 6
ISBN-13: 978 0 340 86918 5

First Published 2004
Impression number   10 9 8 7 6 5 4 3
Year   2009 2008 2007 2006 2005

Copyright © 2004 Kay Clarke, Lynda Francis, Petula Henderson, Cormac McKinney

Produced by Gray Publishing, Tunbridge Wells, Kent
Printed in Italy for Hodder Murray, an imprint of Hodder Education, a member of the Hodder Headline Group, 338 Euston Road, London NW1 3BH

Contents

Introduction   iv

Theme A   Atmosphere and Human Impact   2
Unit One   Weather Patterns and Forecasting   2
Unit Two   Variations in Climate Lead to Different Interactions with
                 Environments   17
Unit Three   Impact of Human Activities upon the Atmosphere and the
                 Environment   25

Theme B   Physical Processes and Challenges   30
Unit One   Crustal Movements and the Impact on People and
                 the Environment (Management Issues)   30
Unit Two   Rivers and River Management   42
Unit Three   Limestone Landscapes and Their Management   56

Theme C   Ecosystems and Sustainability   66
Unit One   Distinct Ecosystems Develop in Response to
                 Climate and Soils   66
Unit Two   Human Interference and Upsetting the Balance of
                 Ecosystems   77
Unit Three   Management of Ecosystems and Sustainable Development   85

Theme D   Population and Resources   90
Unit One   Distribution and Density   90
Unit Two   Population Changes Over Time   94
Unit Three   Population Growth and Sustainability   113

Theme E   Economic Change and Development   122
Unit One   Economic Change Creates New Opportunities   122
Unit Two   The Impact of Global Economic Change   138
Unit Three   Sustainable Development Strategies   147

Theme F   Settlements and Change   170
Unit One   Settlement Development   170
Unit Two   Urban Growth and Change   179
Unit Three   Planning Sustainability for Urban Environments   202

                 All About Learning   214
                 Thinking About What You Have Learned   214

                 World Map and Key to Case Studies   238
                 Ordnance Survey Map Symbols   240
                 Index   241

INTRODUCTION FOR TEACHERS

Geography for CCEA GCSE is intended for 14–16-year-old GCSE geography students in Northern Ireland. The book provides a comprehensive class text for the CCEA GCSE specification and provides students with the knowledge and resources needed to succeed in GCSE geography.

The content is presented following the thematic structure of the CCEA specification, but this does not necessarily imply a proposed teaching order. Teachers can organise the teaching of the content as they consider appropriate and should attempt to emphasise the interrelationship of the different themes. In covering the six themes of the specification, the book addresses the key aspects of the curriculum, Atmosphere and Human Impact, Physical Processes and Challenges, Ecosystems and Sustainability, Population and Resources, Economic Change and Development, and Settlements and Change.

A variety of spatial contexts are used to illustrate the specification content at a range of scales from local to global. The authors have included theoretical information on each topic, definitions of key ideas and appropriate case study material. The final part of the book provides students with a strategy for revision and uses sample answers to exemplify standards in the subject. The command words most frequently used in the examination are explained and techniques for answering different types of questions are suggested.

The book is supplemented with appropriate student activities, including ICT-based tasks for both foundation and higher tier students. Resources suitable for students of all abilities are incorporated in the text, and the book also contains information beyond the scope of the specification to provide teachers with additional resources that could be incorporated into activities designed as extension work to challenge more able students.

When using the book the following should be noted:

■ The websites specified are not always associated with identified activities, but rather, in some cases are included to give students a starting point for further research and to provide sources of further information that could be used to generate additional student activities and discussion. At the time of publication all the websites mentioned are active and accurate.

■ Students need to demonstrate a strong knowledge of place, particularly as exemplified through case studies. The authors have selected certain case studies required by the specification. However, information for all case studies is contained within the book and students could use the information provided to develop their own case study notes. The locations of the case studies are identified, by theme, on the world map on page 238. As case studies are addressed at a range of scales – local/small scale, regional/national scale and international/global scale – an atlas should be used to obtain additional locational information on specific areas as appropriate. Although material has been included for all case studies, this is not intended to be prescriptive, as other examples you may develop will equally be valid in the examination. In fact, the dynamic nature of the subject is such that you will probably wish to develop alternative areas of interest to maintain the vibrancy and topical nature of the subject.

■ Questions are included throughout the book. Some are signposted as being suitable for foundation tier candidates, while others are more appropriate for candidates entered for the higher tier papers.

■ A wide variety of topics from both the physical and human themes of the specification offers students opportunities for the collection, sorting, recording and presentation of primary data, which is an integral part of the compulsory investigative study required by the specification. Possibilities for fieldwork include: local weather, microclimates, studies of river processes and features, ecosystems, soil analysis, population structure and migration, land-use surveys, identifying shopping patterns, mapping spheres of influence, studies of change in residential areas, identifying counterurbanisation and delimiting the central business district (CBD).

INTRODUCTION FOR STUDENTS

The study of geography fosters an understanding of the world and its people. It encourages an appreciation of the relationships between physical and human processes in varied places and environments. Through studying the subject you will have opportunities to develop an understanding of global citizenship, and an awareness of different cultures. I trust that you find the book informative and thought-provoking and hope that you will enjoy following the GCSE course as much as the authors have enjoyed writing this supporting text.

It is my wish that the knowledge and understanding gained, and skills developed, will stimulate your interest and encourage you to continue to develop your appreciation of geography as a dynamic subject, perhaps to continue to an advanced level of study and, I hope, through your lifelong learning.

Kay Clarke
Editor and co-author

ELEMENTS OF THE WEATHER, UNITS AND INSTRUMENTS OF MEASUREMENT

The word **weather** is used to describe the day-to-day changes in the conditions of the **atmosphere** – that is, the weather elements. **Climate** is the average conditions of the weather taken over a long period of time – usually 35 years. The climate of Northern Ireland, for example, is described as mild and damp with few extremely low or high temperatures or large amounts of rain.

Figure 1
Weather takes place in the lowest layer of the atmosphere.

8 km

16 km

EQUATOR

8 km

Elements of the weather

Weather conditions change every day and can vary over short distances even within Northern Ireland, e.g. Belfast can have very different weather to Enniskillen or Coleraine.

Table showing some weather elements, their unit of measurement and the instrument use to record the weather.

Element	Unit	Instrument
Temperature	degree centigrade (°C)	Maximum–Minimum Thermometer
Precipitation	millimetres	Rain Gauge
Wind speed	knots per hour (kph)	Anemometer
Wind direction	8 compass points	Wind Vane
Pressure	millibars (mb)	Barometer

Figure 2
Recording weather.

Maximum–Minimum Thermometer

Barograph

Stevenson's Screen stores instruments

Barometer

Rain Gauge

Wind Vane

Anemometer

activities

1 (a) Measure the weather every day for a week using the instruments shown in Fig. 2. [6]

   (b) Use ICT to construct appropriate graphs. [6]

   (c) Describe the weather patterns shown for the week using the graphs. [10]

2 Explain the difference between weather and climate. [4]

3 (a) Explain why it is necessary to keep thermometers in a Stevenson's Screen? [4]

   (b) Explain how readings are taken from a Maximum–Minimum Thermometer. [4]

   (c) (i) Describe where a Rain Gauge should be located and explain why. [4]

       (ii) Why should a Rain Gauge be checked at the same time each day? [2]

   (d) Estimate today's wind speed and direction. [2]

   (e) Describe how wind speed is measured. [4]

HIGHER TIER

4 Draw a diagram to show what is meant by atmospheric pressure. [4]

5 Explain how pressure is recorded. [4]

Figure 3
The formation of an onshore wind during daytime.

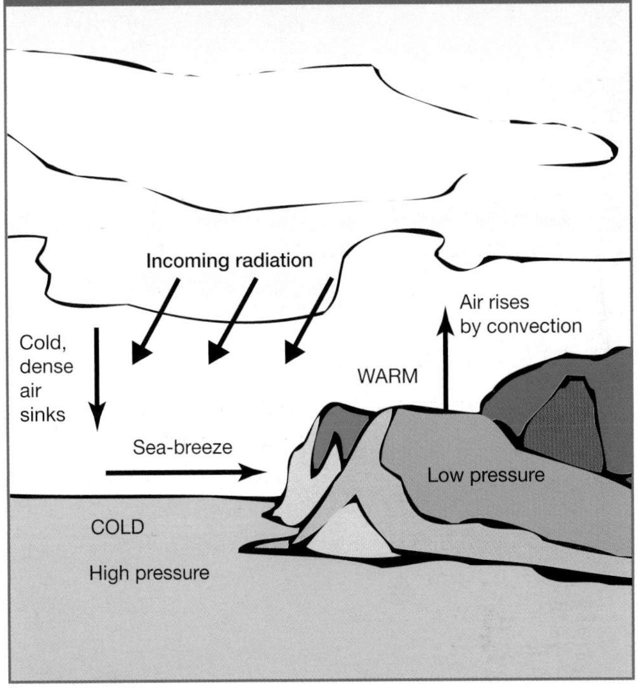

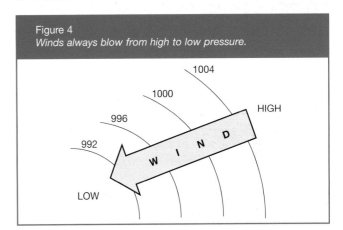

Incoming radiation

Air rises by convection

Cold, dense air sinks

WARM

Sea-breeze

Low pressure

COLD

High pressure

Figure 4
Winds always blow from high to low pressure.

1004
1000
996
992
HIGH
WIND
LOW

Isobars are lines on a weather map. The lines join places of the same pressure. Isobars are numbered at intervals of four millibars.

Pressure and winds

Pressure is the weight of a column of air. The average pressure at sea level is 1012 mb. Winds are movements of air from high to low pressure. Differences in temperature will cause differences in pressure. Heated air expands and becomes less dense. It is light and will rise, so its pressure is low. Cold air contracts and becomes more dense. It is heavy and tends to sink, so its pressure is high.

As air rises, it cools and as air sinks, it warms up. These movements of air result from differences in temperature and pressure. They are illustrated by Fig. 3 for a wind blowing onshore from the sea to the land.

activities

1 (a) Using Fig. 3 describe the formation of a wind blowing from the sea to the land by day or in the summer. [6]

Higher Tier

   (b) Re-draw Fig. 3 to show how a wind blowing from the land to the sea would occur at night or in the winter. [6]

Wind speed

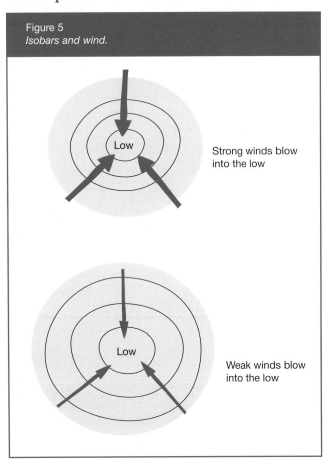

Figure 5
Isobars and wind.

Low

Strong winds blow into the low

Low

Weak winds blow into the low

Wind speeds are high when there is a steep pressure gradient. This is shown on a map by isobars that are close together. A gentle pressure gradient means the wind speeds are low. This is shown on a weather map by isobars that are far apart.

Wind patterns in weather systems

Winds do not blow into areas of low pressure directly, but at an angle. This is because the Earth rotates on its axis once every 24 hours, which means that winds blow to the right of their path of movement in the northern hemisphere. As a result winds blow at an angle across the isobars into areas of low pressure, which means that winds blow anti-clockwise and into centres of low pressure and they blow clockwise and out of areas of high pressure.

1   On a copy of Fig. 7:

   (a) Number the isobar at X.   [1]

   (b) Mark HIGH in the area of highest pressure and LOW in the area of lowest pressure   [2]

   (c) Add arrows to indicate the wind direction over Ireland.   [3]

HIGHER TIER

   (d) Complete the isobar pattern by adding the isobars at intervals of 4 millibars and numbering each isobar.   [4]

   (e) Add 10 arrows to show wind direction. Spread the arrows across the map and draw them the same length as the arrow in the key.   [4]

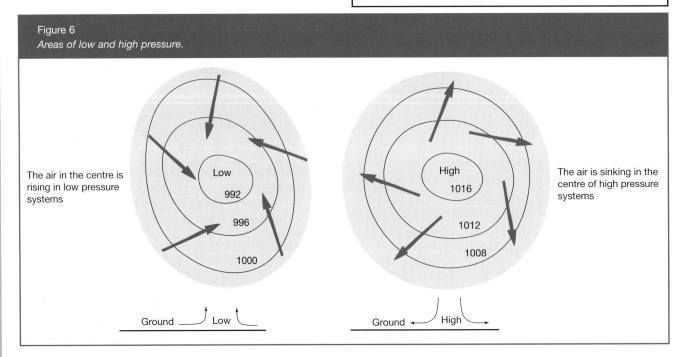

Figure 6
Areas of low and high pressure.

The air in the centre is rising in low pressure systems

Low
992
996
1000

The air is sinking in the centre of high pressure systems

High
1016
1012
1008

Ground ___ Low

Ground ← High →

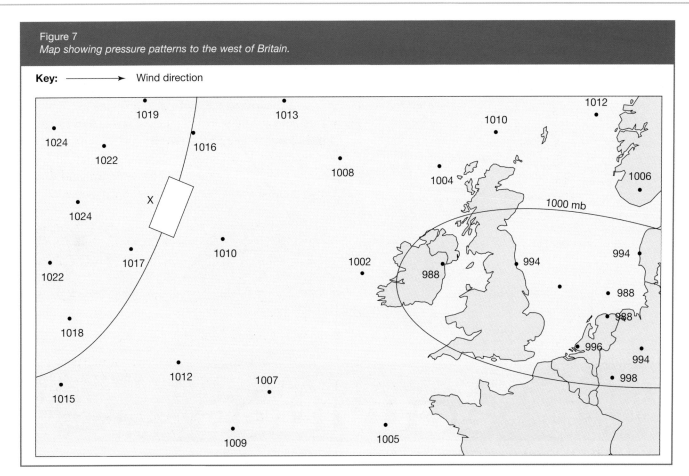

Figure 7
Map showing pressure patterns to the west of Britain.

Key: ⟶ Wind direction

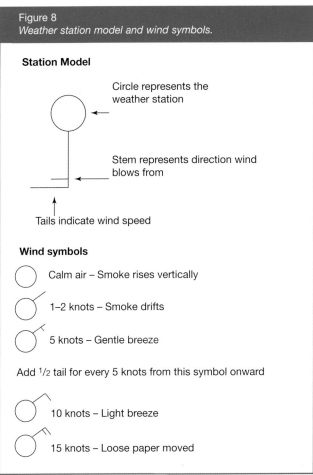

Figure 8
Weather station model and wind symbols.

Station Model

Circle represents the weather station

Stem represents direction wind blows from

Tails indicate wind speed

Wind symbols

Calm air – Smoke rises vertically

1–2 knots – Smoke drifts

5 knots – Gentle breeze

Add ¹/₂ tail for every 5 knots from this symbol onward

10 knots – Light breeze

15 knots – Loose paper moved

Precipitation

The term precipitation includes all types of moisture in the atmosphere from rainfall to hail and snow to mist and fog. Rain occurs when moist air cools when the air is forced to rise. The damp air containing water vapour rises and cools; the water vapour it is holding condenses and forms water droplets. These can be held in the air as clouds or, when the water droplets are large and heavy enough, they can fall as rain. Clouds are masses of water droplets of different shapes and form at different heights above the ground.

There are different ways in which the air is made to rise – it can rise over mountains (relief rainfall) or at **fronts** in **depressions** (cyclonic rainfall). Warm moist air can also be forced to rise due to great heating of the ground (convectional rainfall); this is common in areas of Tropical Rainforest (see Theme C).

Relief rainfall

Relief rainfall forms when moist air is forced to rise over a mountain. The air cools as it rises and the water vapour condenses to form cloud and rain at the top of the mountain. After it passes over the mountain the air sinks and warms up so that there is

Figure 9
Relief rainfall.

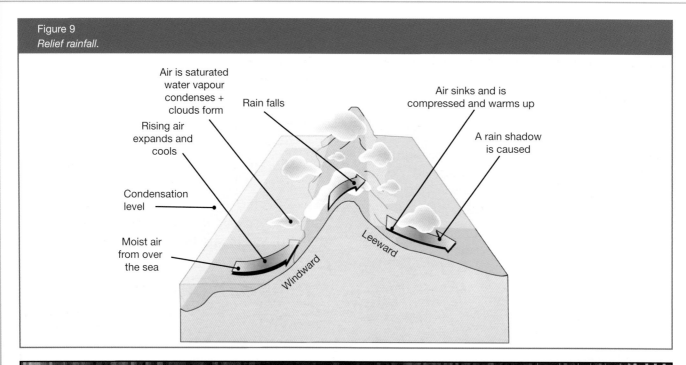

Air is saturated
water vapour
condenses +
clouds form

Rain falls

Air sinks and is
compressed and warms up

Rising air
expands and
cools

A rain shadow
is caused

Condensation
level

Moist air
from over
the sea

Leeward

Windward

activities

1 (a) Sort the following phrases into the
correct order to show the stages
in the formation of relief rainfall.   [5]

Clouds form and rain falls on the
windward side of the mountain.

Air sinks and is compressed and
warms up.

Damp air blows in from the sea.

Air is forced to rise over the mountain

The area on the leeward side of the
mountain is dry.

Air cools and the water vapour
condenses

   (b) Complete the four tasks on Thinking
Skills 1 opposite.

HIGHER TIER

   (c) Compare and contrast relief and
frontal rainfall.   [8]

Thinking Skills I – Weather

1. temperature	2. cyclone	3. thermometer
4. cirrus	5. gale	6. cumulonimbus
7. typhoon	8. wind vane	9. breeze
10. rainfall	11. snow	12. nimbus
13. calm	14. storm	15. rain gauge
16. cumulus	17. hurricane	18. wind direction
19. cloudy	20. barometer	

Task 1
Each of the numbers in the sets of four relates to the topic
weather. Can you work out with your partner which is the
Odd One Out and what connects the other three?

Set A	3	8	15	18
Set B	1	10	3	18
Set C	4	19	12	16
Set D	12	5	13	9
Set E	7	8	2	17

Task 2
Still with your partner, can you find *one more* from the list to add
to each of the sets above so that *four* items have things in
common, but the *Odd One Out* remains the same? Think about
why you have chosen each one.

Task 3
Now it's your turn to design some sets to try out on your partner!
Choose three numbers that you think have something in common
with each other and one that you think has nothing to do with the
other two. Get your partner to find the *Odd One Out*, then do one
of theirs. Try a few each, but remember to be reasonable.

Task 4
Can you organise all the words into groups? You are allowed to
create between 3 and 6 groups and each group must be given a
descriptive heading that unites the words in the group. Try not to
have any left over. Be prepared to rethink as you go along.

Source: *Thinking Through Geography Material*, N.I. Education
Boards (Nov. 1999)

no rain on this side. This type of rainfall is common in the north and west of the British Isles. In the uplands to the west of Northern Ireland there is about 1750 mm of rain per year but the east only receives 1000 mm of rain per year.

AIR MASSES

The British Isles is affected by five main air masses and this makes the weather very changeable. An **air mass** is a large body of air with similar temperature and moisture characteristics all the way through it.

The characteristics of the air mass depend on where it comes from or the source region:

- A MARITIME air mass picks up moisture from the sea surface and so brings wet weather to the British Isles;

- A CONTINENTAL air mass is dry because it forms over land surfaces;

- A POLAR air mass comes from a northerly direction and so brings cold temperatures;

- A TROPICAL air mass comes from a southerly direction and so brings warm temperatures;

- An ARCTIC air mass comes from a northerly direction and brings cold, snowy weather in winter and cool, damp weather in summer.

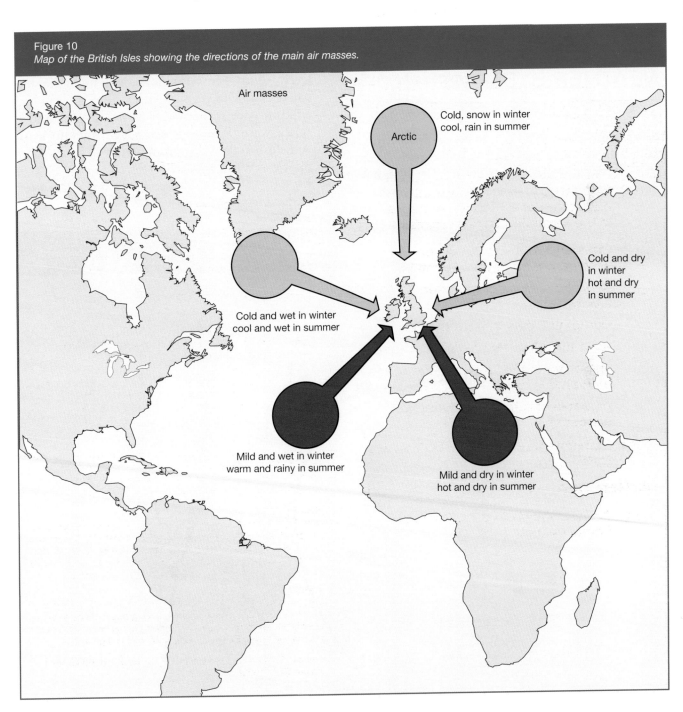

Figure 10
Map of the British Isles showing the directions of the main air masses.

Air masses

Arctic — Cold, snow in winter cool, rain in summer

Cold and wet in winter cool and wet in summer

Cold and dry in winter hot and dry in summer

Mild and wet in winter warm and rainy in summer

Mild and dry in winter hot and dry in summer

Where air masses meet, a front is formed. A **front** separates warm and cold air masses. The Tropical Maritime air mass is usually found between the warm and cold fronts (in the warm sector) of depressions.

activities

On an outline world map, copy the air masses shown on Fig. 10 and complete by labelling the following air masses in the correct locations:

Polar Maritime   Tropical Maritime

Polar Continental   Tropical Continental.

WEATHER SYSTEMS

Depressions

Depressions are systems of low pressure; they are like whirlpools of air that develop in the main stream of air movement which comes towards the British Isles from the west. They are areas of low pressure which generally move towards the east. The winds blow anticlockwise and into the centre of the low. The air rises in the centre.

On weather maps a depression has a circular pattern of isobars with the lowest pressure in the centre and the winds blowing into the centre. Depressions can be hundreds of kilometres wide. Depressions have fronts because they form when a warm tropical air mass meets a cold polar air mass. A front divides the two air masses. The warm front and the cold front are separated by a wedge of warm tropical air called the warm sector.

The sequence from an observer's point of view

■ Ahead of the Warm Front

As a depression approaches, a person at ground level will first see cirrus cloud, high up in the sky. There is no rain yet but temperatures are cool and winds may be strong and from the east or south. As the warm front approaches, the cloud thickens and close to the warm front there will be rain and drizzle as warm air is being forced to rise. Pressure is high but decreases towards the centre of the low pressure.

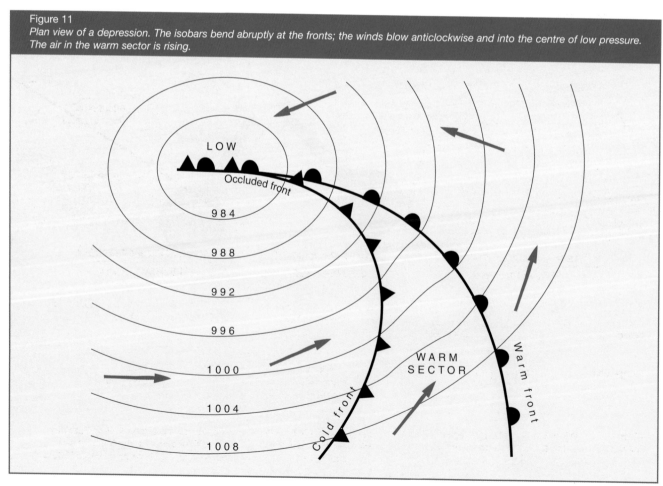

Figure 11
Plan view of a depression. The isobars bend abruptly at the fronts; the winds blow anticlockwise and into the centre of low pressure. The air in the warm sector is rising.

Figure 12
Cross-section of a depression.

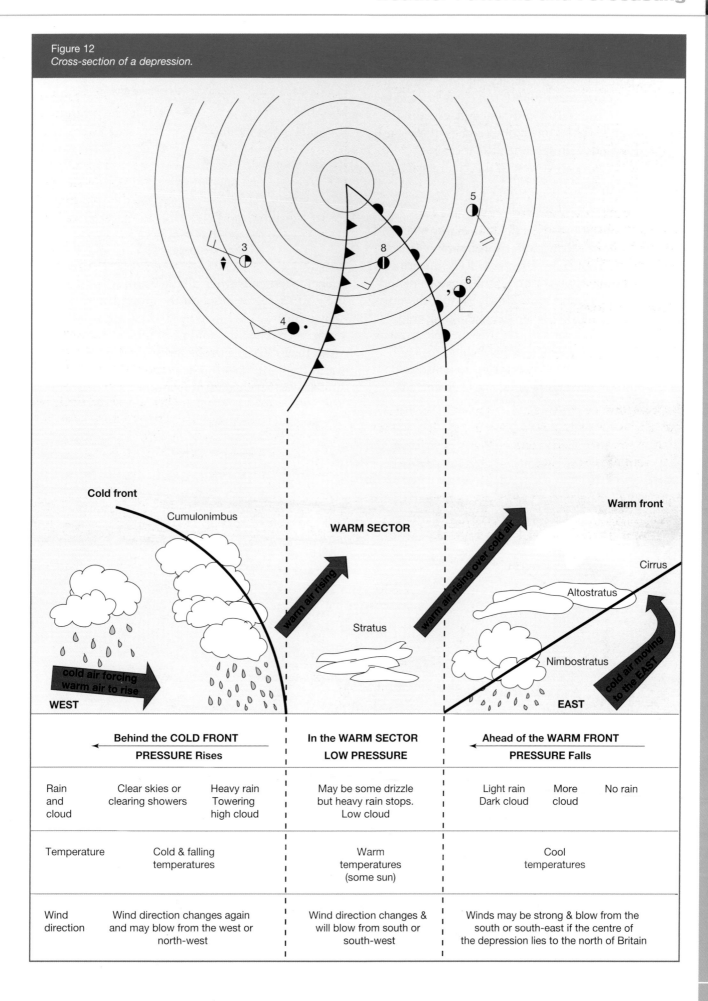

	Behind the COLD FRONT		In the WARM SECTOR	Ahead of the WARM FRONT		
	← PRESSURE Rises		LOW PRESSURE	← PRESSURE Falls		
Rain and cloud	Clear skies or clearing showers	Heavy rain Towering high cloud	May be some drizzle but heavy rain stops. Low cloud	Light rain Dark cloud	More cloud	No rain
Temperature	Cold & falling temperatures		Warm temperatures (some sun)	Cool temperatures		
Wind direction	Wind direction changes again and may blow from the west or north-west		Wind direction changes & will blow from south or south-west	Winds may be strong & blow from the south or south-east if the centre of the depression lies to the north of Britain		

- In the Warm Sector

  Here temperatures increase in the warm tropical air. There is low stratus cloud but it is mainly dry; this is because water vapour can easily be held in the warm, tropical air without condensation taking place. Wind direction becomes more south-westerly and wind speed usually increases. Pressure values are lowest in this central part of the depression.

- Behind the Cold Front

  As the cold front passes, the temperatures fall and the winds will change direction and blow from the north-west. The observer will see towering high cumulonimbus clouds at the cold front and there will be heavy rain. This is because the warm air is rising quickly at the steeply sloping cold front. Pressure starts to rise after the cold front as the depression passes. Further behind the cold front are scattered showers from some isolated cumulus clouds; the wind speeds become lighter.

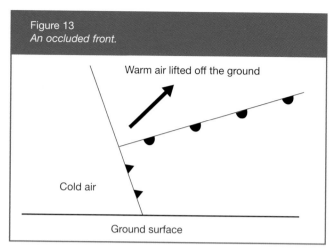

Figure 13
An occluded front.

A depression ends when all the warm tropical air in the warm sector is lifted off the ground. This is shown on a weather map by an occluded front. This can happen when the cold front moves eastwards faster than the warm front and so cold air pushes the warm front off the ground. An occluded front brings similar weather to a warm front.

Figure 14
The sequence of weather as a depression passes.

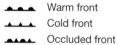
Warm front
Cold front
Occluded front

1. A warm front approaches. Skies change from clear, to having high wispy clouds, to developing thicker, lower clouds. Eventually it starts to drizzle.

2. The warm front passes over. Drizzle is replaced by steady rain.
   The temperature rises in the warm sector. The sky remains grey and overcast. Rain or drizzle continues. The wind blows from the south-west.

3. The cold front has passed over Northern Ireland resulting in heavy rain and gusty conditions. Temperatures drop as the front passes over. The wind swings around to the north-west. Behind the cold front, the sky clears. The weather now brings sunny intervals with heavy showers.

activities

1 (a) Using Fig. 14 describe how the temperatures change at Belfast through the day. [3]

(b) Explain why temperatures change as the depression passes. [6]

(c) Describe how the wind direction changes at Derry through the day. [3]

(d) Explain why the rainfall amount changes during the day. [4]

HIGHER TIER

2 Describe how the changing weather throughout the day would affect each of the following groups of people.

- Tourists in Newcastle, Co. Down.
- Hill walkers in the Sperrin Mountains.
- Travellers taking the Larne to Cairnryan ferry. [9]

Anticyclones

Anticyclones are systems of high pressure. In the centre the air is sinking slowly from great heights; as the air sinks it swirls in a clockwise direction and spreads out at the surface. The sinking air is compressed and warms up as it nears the ground;

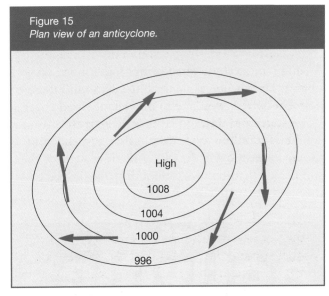

Figure 15
Plan view of an anticyclone.

High
1008
1004
1000
996

this means the air can hold more water vapour without condensation taking place and so clouds do not form and it is less likely to rain. This means anticyclones are associated with dry, bright weather.

In an anticyclone the isobars are spaced well apart and so the pressure gradient is gentle; this means the wind speeds are low and there may even be calm conditions with no wind in the centre of the high pressure. An anticyclone has no fronts and moves very slowly so the weather conditions may not change very much as this system passes across the British Isles.

A summer anticyclone brings cloudless skies and bright sunshine and high temperatures during the day. At night temperatures can fall due to rapid cooling caused by heat escaping through radiation

Figure 16
The weather in a summer anticyclone and a winter anticyclone.

into the atmosphere when there are no clouds. This mist is usually easily evaporated by the strong sunshine in summer.

A winter anticyclone often brings fog or mist; these form at night when rapid cooling occurs and there is heat lost by radiation due to the lack of cloud cover. Water vapour in the cold layer of air near the ground condenses and the water droplets are suspended in the air as mist or fog. In winter the low angle of the sun means that the rays cannot disperse the fog or mist and it may persist all day.

activities

1 Describe the weather conditions shown in the two photographs in Fig. 16. [4]

2 Explain why anticyclones bring sunny daytime weather but cold weather at night in winter. [5]

HIGHER TIER

3 Design a diagram which shows the movement of air in an anticyclone. Use the information provided in the description of this system. [4]

web link & extra resources

www.onlineweather.com

www.meto.gov.uk

www.bbc.co.uk/ni/education

http://weather.noga.gov/weather

SYNOPTIC CHARTS

Synoptic Charts are maps which summarise the weather conditions at a particular point in time for an area. They record the weather using a set of symbols and show the fronts of a depression and the variation in the pressure of the air using isobars.

Satellite images

A satellite image is a photograph taken from space and sent back to Earth; it can show the cloud formations and the pattern of clouds at fronts in depressions or the clear skies associated with high pressure areas.

Weather forecasting

A **weather forecast** is made using computers and the records of past weather patterns to predict current weather. Forecasters use data collected from

Figure 17
Weather symbols used on weather maps.

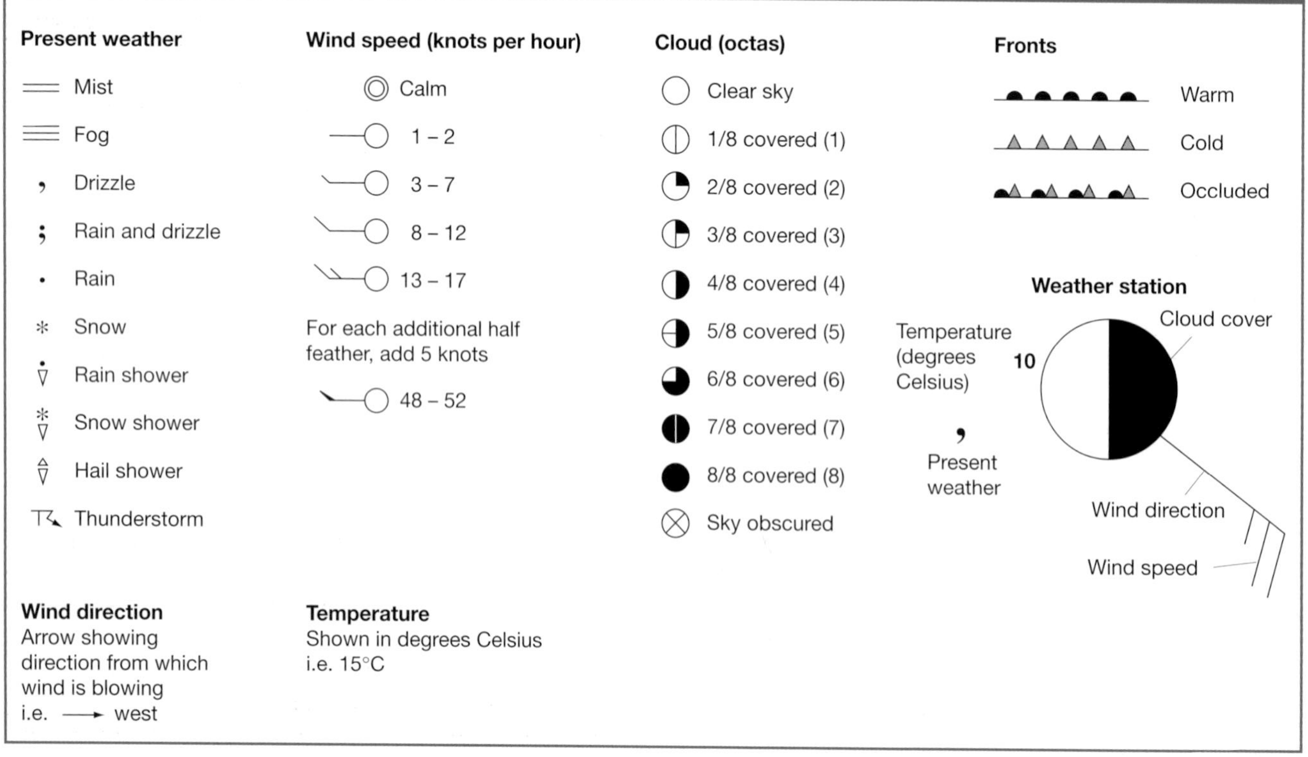
Present weather
Mist
Fog
Drizzle
Rain and drizzle
Rain
Snow
Rain shower
Snow shower
Hail shower
Thunderstorm
Wind speed (knots per hour)
Calm
1 – 2
3 – 7
8 – 12
13 – 17
For each additional half feather, add 5 knots
48 – 52
Cloud (octas)
Clear sky
1/8 covered (1)
2/8 covered (2)
3/8 covered (3)
4/8 covered (4)
5/8 covered (5)
6/8 covered (6)
7/8 covered (7)
8/8 covered (8)
Sky obscured
Fronts
Warm
Cold
Occluded
Weather station
Cloud cover
Temperature (degrees Celsius)
10
Present weather
Wind direction
Wind speed
Wind direction
Arrow showing direction from which wind is blowing
i.e. west
Temperature
Shown in degrees Celsius
i.e. 15°C

Figure 18
Inputs and outputs of weather data.

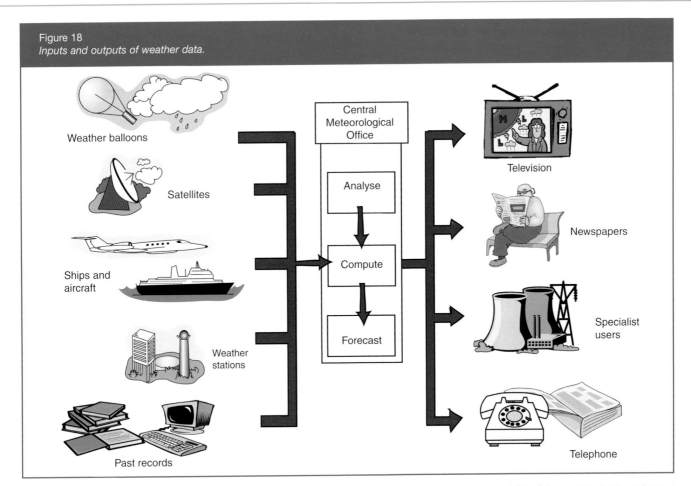

many sources to produce a synoptic chart which shows the predicted weather conditions.

Short-range forecasting

This is usually for a period of 4–5 days ahead. Computers forecast the weather from the accurate data collected from many sources. Weather patterns continually change and vary with both **altitude** (height) and space. This means accurate predictions can only be made for a few days ahead.

Long-range forecasting

These forecasts depend on the movement of air masses which is affected by many variables, so accuracy of prediction falls when forecasts are more than 5 days ahead. Another difficulty is that forecasters must predict how the weather will vary over very large areas and also how the conditions of the weather will change vertically with height.

User forecasts

A weather forecast describes the expected weather conditions. Different groups of people rely on these for their activities. For example, insurance companies require data on the likelihood of flooding in a particular area so that they can set a premium based on the risk to homeowners. Builders and

activities

CLIMATE	The normal or average conditions for the time of year
MOVEMENT of AIR MASSES, FRONTS and PRESSURE SYSTEMS	In which directions will they move or change?
TIME OF DAY	The height of the sun, whether it is day or night
PRESSURE VALUE	High or low
MOISTURE	Amount and type of precipitation (rain, hail, mist, snow, etc.)

1 (a) Place the forecasting factors above into their order of importance with 1 being the most important in forecasting the weather.   [4]

HIGHER TIER

(b) Prepare a weather forecast for the next 12 hours for your local area; predict the temperature, wind speed and direction, cloud cover and precipitation, and the pressure. Use the table of factors to help you.   [10]

construction workers may also need to know the weather so that they can plan their work on a site: if it is expected to rain heavily, they may decide to postpone work outside such as digging foundations and work on indoor jobs such as flooring or plastering instead.

Figure 19
Groups of people who rely heavily on weather forecasts.

Transport operators · Farmers · Event organisers · People who use weather forecasts · Cyclists · Hospital managers · Sailors · Sport players

activities

Choose two of the groups in Fig. 19 and suggest why they might require a weather forecast [6]

activities

1 In Fig. 20, Table (a) is for Dublin. Using both the satellite image in Fig. 21 and the synoptic chart in Fig. 20, explain why this weather forecast accurately describes the conditions in Dublin. [6]

HIGHER TIER

2 (a) Using Fig. 21, match the weather forecast in the other tables (b, c and d) to the cities of Oslo, Alicante and Rome. [3]

(b) Justify your choice in each case, using both the satellite image and the synoptic chart. [9]

Figure 20
Tables (a–d) showing weather forecasts for four cities in Europe.

a) Dublin 5 day forecast

	Tuesday 18th February		Wednesday 19th February		Thursday 20th February		Friday 21st February		Saturday 22nd February
	Day Max	Night Min	Day Max	Night Min	Day Max	Night Min	Day Max	Night Min	Day Max
	6°C	−1°C	6°C	−2°C	8°C	1°C	7°C	3°C	8°C

b)

	Tuesday 18th February		Wednesday 19th February		Thursday 20th February		Friday 21st February		Saturday 22nd February
	Day Max	Night Min	Day Max	Night Min	Day Max	Night Min	Day Max	Night Min	Day Max
	−4°C	−12°C	−3°C	−12°C	−3°C	−16°C	−4°C	−10°C	−4°C

c)

	Tuesday 18th February		Wednesday 19th February		Thursday 20th February		Friday 21st February		Saturday 22nd February
	Day Max	Night Min	Day Max	Night Min	Day Max	Night Min	Day Max	Night Min	Day Max
	13°C	7°C	14°C	7°C	17°C	6°C	16°C	9°C	16°C

d)

	Tuesday 18th February		Wednesday 19th February		Thursday 20th February		Friday 21st February		Saturday 22nd February
	Day Max	Night Min	Day Max	Night Min	Day Max	Night Min	Day Max	Night Min	Day Max
	11°C	−1°C	11°C	1°C	13°C	3°C	12°C	1°C	13°C

Figure 21
Synoptic chart of an anticyclone and its satellite image.

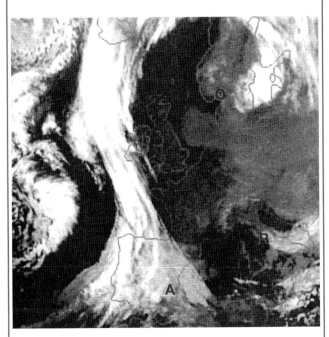

Alicante (A), Dublin (D), Oslo (O) and Rome (R)

activities

1   Complete the four tasks on Thinking
    Skills 2.

Thinking Skills 2 – Weather

1. rising air	2. rain
3. isobars	4. hot weather
5. increasing winds	6. rain gauge
7. drizzle	8. depression
9. sinking air	10. fronts
11. pressure rising	12. anticyclone
13. warm front	14. pressure
15. hail	16. barometer
17. cold front	18. snow
19. temperature rising	20. warm sector
21. stratus clouds	22. calm winds
23. anticlockwise winds	24. cold air
25. clouds	26. frost
27. clockwise winds	28. clear skies
29. precipitation	30. high temperatures
31. temperature falling	32. relief rain

Task 1

Each of the numbers in the sets of four relates to the topic
weather. Can you work out with your partner which is the
Odd One Out and what connects the other three?

Set A	30	22	2	28
Set B	13	25	9	5
Set C	3	14	16	6
Set D	25	2	15	18
Set E	11	17	31	1
Set F	20	21	7	24
Set G	23	32	8	10
Set H	12	29	27	22

Task 2

Still with your partner, can you find *one more* from the list
to add to each of the sets above so that *four* items have
things in common, but the *Odd One Out* remains the
same? Think about why you have chosen each one.

Task 3

Now it's your turn to design some sets to try out on your
partner! Choose three numbers that you think have something
in common with each other and one that you think has nothing
to do with the other two. Get your partner to find the *Odd One
Out*, then do one of theirs. Try a few each, but remember to
be reasonable.

Task 4

Can you organise all the words into groups? You are allowed
to create between three and six groups, and each group must
be given a descriptive heading that unites the words in the
group. Try not to have any left over. Be prepared to rethink
as you go along.

Source: *Thinking Through Geography Material*, N.I. Education
Boards (Nov. 1999)

2   Complete the ICT exercise.

ICT Exercise

A typical October depression – an exercise using ICT

Passage of a depression

Date	Time	Rainfall (mm)	Temp. (°C)
29/10/00	10.00	0.0	10.0
29/10/00	11.00	0.0	10.8
29/10/00	12.00	0.0	11.6
29/10/00	13.00	0.0	11.6
29/10/00	14.00	0.0	11.8
29/10/00	15.00	0.0	11.6
29/10/00	16.00	0.0	10.5
29/10/00	17.00	0.2	10.8
29/10/00	18.00	1.6	11.4
29/10/00	19.00	2.0	12.1
29/10/00	20.00	2.4	11.7
29/10/00	21.00	2.4	11.7
29/10/00	22.00	1.6	13.2
29/10/00	23.00	0.8	11.4
30/10/00	00.00	5.4	13.1
30/10/00	01.00	2.4	13.3
30/10/00	02.00	2.4	13.2
30/10/00	03.00	1.6	13.1
30/10/00	04.00	1.6	13.0
30/10/00	05.00	2.0	13.1
30/10/00	06.00	1.4	12.8
30/10/00	07.00	2.8	9.8
30/10/00	08.00	5.0	10.2
30/10/00	09.00	1.0	8.9
30/10/00	10.00	0.0	8.4
30/10/00	11.00	0.0	6.6
30/10/00	12.00	0.2	7.6
30/10/00	13.00	0.4	9.9
30/10/00	14.00	0.2	10.8
30/10/00	15.00	0.0	11.3
30/10/00	16.00	0.0	11.0
30/10/00	17.00	0.0	10.5

Source: *Cut, Paste & Surf* by Philip Webster (Nelson Thornes, 2002)

Activities

1 Use the data provided to create a spreadsheet. [6]

2 Draw appropriate graphs to show how rainfall and temperature change as a depression passes (bar + line graphs). [6]

3 Add suitable titles and axes labels to each graph. [3]

4 Write a paragraph to describe how the weather changes as a depression passes. [6]

HIGHER TIER

5 Explain why these changes occurred, using your knowledge of the depression weather system. [6]

UNIT TWO
Variations in Climate Lead to Different Interactions with Environments

CAUSES OF VARIATION IN CLIMATE BETWEEN PLACES

Altitude

Altitude affects temperature which falls by 1°C for every 100 metres of height, e.g. the top of Slieve Donard at 852 metres will by cooler by 8.5°C than at sea level in Newcastle, Co. Down. Higher areas usually have more rainfall because air that rises over mountains will cool and the water vapour will condense resulting in cloud and rain.

Latitude

Latitude helps to explain why temperatures are normally warmer in the south of the British Isles than in the north. The sun's rays are more concentrated towards the equator and they hit the Earth more directly overhead; they also pass through a smaller depth of atmosphere at the equator so there is less chance of energy being deflected by dust and other particles. This results in areas further south receiving more solar radiation and so the south is warmer than the north in Britain.

Continentality

Distance from the sea or **continentality** affects both temperature and rainfall. It takes more energy to heat the sea, so the sea heats up and cools more slowly than landmasses. This means that places near the sea are relatively cooler in summer and warmer in winter than places inland. In winter, places inland are much colder than places near the sea.

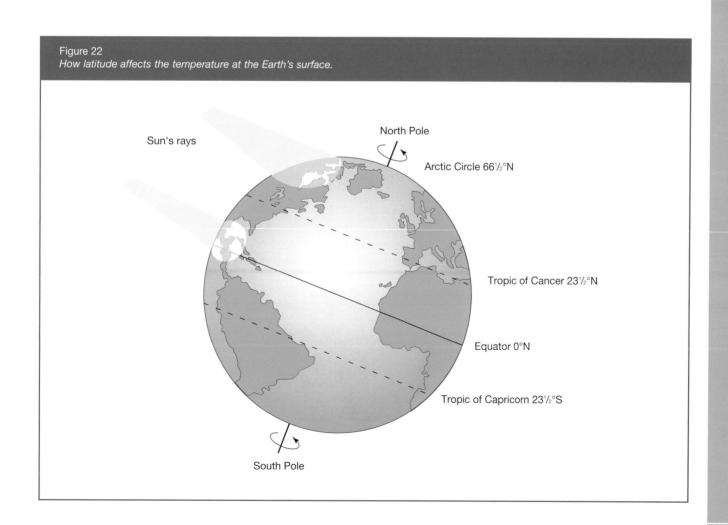

Figure 22
How latitude affects the temperature at the Earth's surface.

Sun's rays

North Pole

Arctic Circle 66½°N

Tropic of Cancer 23½°N

Equator 0°N

Tropic of Capricorn 23½°S

South Pole

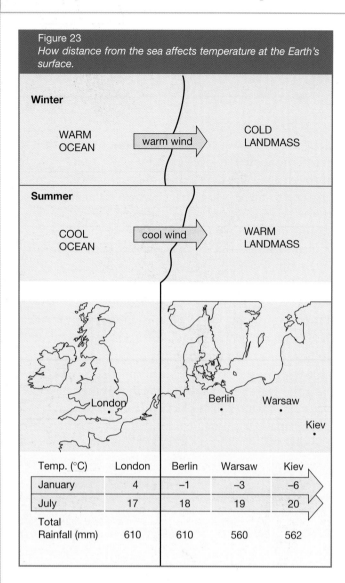

Figure 23
How distance from the sea affects temperature at the Earth's surface.

Temp. (°C)	London	Berlin	Warsaw	Kiev
January	4	–1	–3	–6
July	17	18	19	20
Total Rainfall (mm)	610	610	560	562

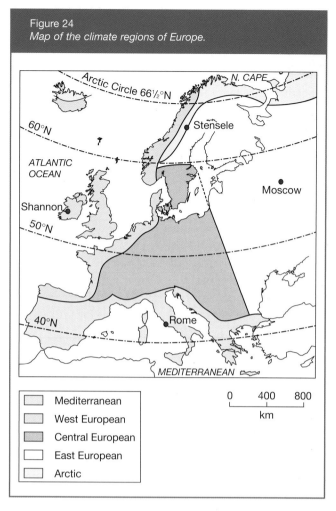

Figure 24
Map of the climate regions of Europe.

Mediterranean
West European
Central European
East European
Arctic

Places inland are drier than coastal places because winds blowing over the sea evaporate moisture from the sea surface and bring rain to coastal places.

Prevailing wind

The **prevailing wind** is the most frequently occurring wind blowing towards a place. Over Ireland the prevailing wind is westerly from over the Atlantic Ocean. This means that moisture is brought towards the British Isles. As the ocean is relatively warm in winter, the climate of the British Isles is mild and wet in winter, and cool and wet in summer.

How these factors affect the climate of the continent of Europe

Although Europe is a relatively small continent in area, it has five different climatic regions. The height of the land ranges from below sea level in the Netherlands to over 4800 metres at Mt Blanc in the French Alps. Europe stretches from 35 to 71°N so the angle of the overhead sun will vary from the north of Norway south to the Mediterranean coastline. Some places in Europe are a very long way inland and so are far from the influence of the sea. The prevailing winds over most of Europe are from the west across the Atlantic Ocean, but eastern Europe experiences cold dry air in winter and Mediterranean areas receive southerly winds from North Africa in summer.

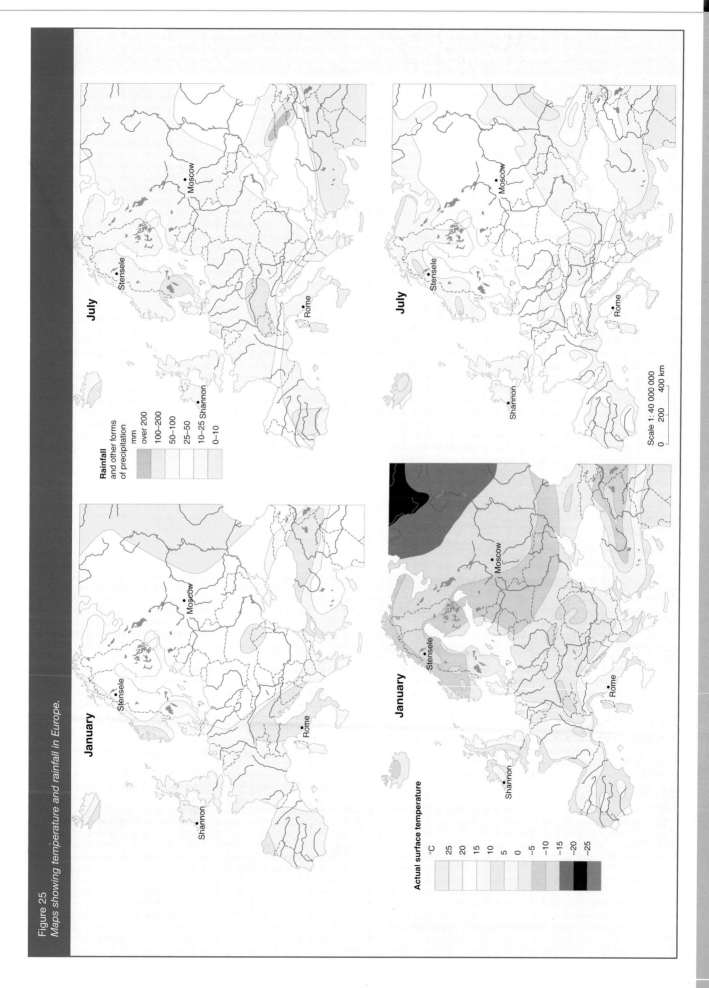

Figure 25
Maps showing temperature and rainfall in Europe.

July

January

July

January

Rainfall
and other forms
of precipitation

| mm |
| over 200 |
| 100–200 |
| 50–100 |
| 25–50 |
| 10–25 |
| 0–10 |

Moscow

Stensele

Rome

Shannon

Scale 1: 40 000 000
0   200   400 km

Actual surface temperature

°C
25
20
15
10
5
0
–5
–10
–15
–20
–25

Figure 26
Climate graphs for four locations in Europe.

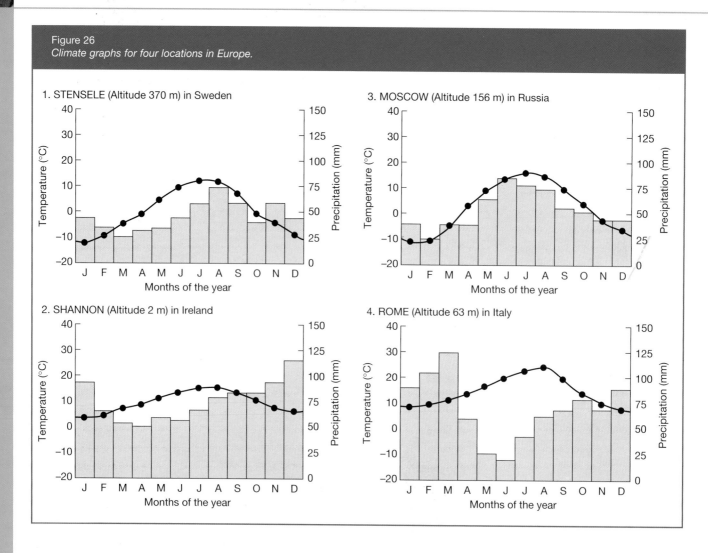

1. STENSELE (Altitude 370 m) in Sweden
2. SHANNON (Altitude 2 m) in Ireland
3. MOSCOW (Altitude 156 m) in Russia
4. ROME (Altitude 63 m) in Italy

activities

1   Copy and complete the table below by ranking the four named places on Fig. 26. [12]

	Stensele	Shannon	Moscow	Rome
Warmest in summer				
Mildest in winter				
Highest difference in temperature				
Most rainfall				

2   Which of the four named places has the following climate characteristics?

   (a) warm, wet winters and hot, dry summers.

   (b) cold, dry winters and cool, wet summers. [2]

3   Draw a graph to show the change in the temperatures and rainfall across Europe from London to Kiev using Fig. 23. [4]

HIGHER TIER

4   Explain why London and Kiev have such different climates. [6]

5   Explain why Rome's climate is so different from Stensele's. [6]

POSITIVE AND NEGATIVE IMPACT OF CLIMATE ON FARMING

CASE STUDY

THE EXAMPLE OF EAST ANGLIA IN THE UK

Figure 27
Farming in East Anglia.

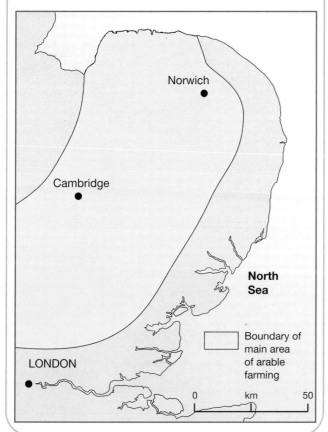

North Sea

Norwich

Cambridge

LONDON

☐ Boundary of main area of arable farming

0   km   50

East Anglia is an area of arable farming (crop growing). Cereal crops such as wheat, barley and sugar beet are cultivated, as well as potatoes and vegetables such as peas and beans. There are several reasons for this type of agriculture being carried out in this part of England: the land is almost flat, which suits the use of large machines, and the soils are easy to work and rich in organic matter so they are fertile; but the most important reason for the cultivation of cereals in this region is that the climate suits these crops.

Climate of East Anglia (Norwich area)

	J	F	M	A	M	J	J	A	S	O	N	D
Temp (°C)	4	4	6	8	12	15	17	18	14	11	7	4
Rainfall (mm)	61	46	38	48	43	43	66	53	58	64	69	61

This climate has a positive impact and suits arable farming

■ This part of eastern England is the driest part of the UK and receives lower rainfall than the west (rainfall total is below 700 mm per year). The low rainfall suits the growth of cereal crops.

■ There are long hours of sun in this region to the south of the UK (over 1500 hours of sunshine per year) which results in less cloud cover so allowing the crops to ripen (6.5 hours of sun per day in July).

■ Summers are warm with average temperatures over 15°C which helps the crops to ripen early.

■ The autumn season tends to be dry (August and September) which allows farmers to harvest the crops easily.

■ Winters are cold with hard frosts (the average temperature in January is only 4°C). The frosts in winter help to break up the soil, ready for ploughing in spring and the cold winters help to kill pests which would otherwise spread disease.

This climate has a negative impact and causes problems for farmers

■ In some years the climate is too dry during the summer months for cereal crops or vegetables such as peas and beans to grow successfully.

continues

continues

To solve this problem additional water must be added by spraying crops with water or through channels around the fields; adding water to dry land is called irrigation. However, this can also create its own problems as providing irrigation water is expensive and, if irrigation continues for long periods, the soil will dry out so much that salts can move up to the surface; saline or salty soils prevent crop growth.

■ In some years there are very late frosts when the air temperatures fall below 0°C. This can kill off the young plants before they are fully

continues

grown or kill off the vegetables before they are ready for harvest.

To prevent this, crops have to be protected using heaters in the fields or by using cloches to cover the vegetable crops.

East Anglia is the most important farming region in the UK, but only a small percentage of the region's people are employed in farming. This is because this type of farming is very intensive and commercial; machines and computers do the work and only a small labour force is needed to work each farm. Farms here are very large – about 200 hectares.

activities

1 Draw a graph to show both the temperature and rainfall of this climate [8]

2 State the meaning of the term arable farming. [2]

3 State the total rainfall. [2]

4 (a) Describe the rainfall amounts in August and September. [3]

(b) State the work done on the farm during these months using the diagram below. Explain why this work can be done at this time. [4]

5 Explain why having long hours of sunshine in summer is an advantage to this type of farming. [3]

HIGHER TIER

6 (a) Describe **one** negative impact the climate of East Anglia has on farming and how it is overcome. [5]

(b) Describe **two** ways in which climate suits the type of farming in East Anglia. [6]

7 Draw an annotated sketch map to show how climate influences an area of arable farming in East Anglia. [8]

					Months							
---	O	N	D	J	F	M	A	M	J	J	A	S
PLOUGHING												
SOWING & DRILLING				wheat		barley		peas beet				
FERTILISERS												
HOEING												
HARVESTING									peas		wheat barley	
MAINTENANCE		hedging & ditching machine										

CASE STUDY

THE EXAMPLE OF FARMING IN THE MEDITERRANEAN

Figure 28
Areas in southern Spain where olive trees and citrus fruits are grown.

This farming region between the Sierra Nevada Mountains and the Mediterranean coast to the east of Malaga can grow a wide variety of crops which suit the climate with its hot sunny summers and warm wet winters. The traditional crops of the Mediterranean region are olives and grapes. These crops are grown on the gentle valley slopes facing south to make the most of the high angle of the sun in summer, e.g. in areas around Orgiva. Farms are small, often only 5 hectares. Sheep and goats are sometimes kept.

Olive trees have grey-green leaves and are evergreen; they produce their purple-coloured fruit in early winter. The olives are pressed to make olive oil; olive trees produce about 20 kg of olive oil each when the olives are pressed. Grapes are grown on vines and most are used to make wine.

In the past farmers grew only crops which suited the long dry summers. Now more intensive farming of citrus fruit (oranges and lemons), soft fruit (peaches) and vegetables is possible with the use of irrigation water.

Climate of the Mediterranean region (Orgiva area)

	J	F	M	A	M	J	J	A	S	O	N	D
Temp. (°C)	10	12	13	15	18	20	23	25	23	17	14	10
Rainfall (mm)	38	34	52	32	30	16	5	25	54	76	48	40

This climate has a positive impact and suits olive, grape and citrus farming

- Hot sunny summers with high temperatures of over 20°C help the fruit to ripen; June to August is very sunny and this helps to increase the sugar content of the grapes.

- Short cold winters suit the growth of vines because this strengthens the roots and helps the growth of the vine stems and also to kill pests.

- Most rainfall occurs in late winter and early spring, and this suits the olive trees as the moisture helps to swell the fruit.

- The temperatures do not fall below 10°C and this suits the citrus fruit which do not cope well with cold temperatures.

This climate has a negative impact and causes problems for farmers

- Long dry summers with no rain and high rates of evaporation create drought conditions and many crops cannot grow without the artificial addition of water. However, there is great potential for farming if irrigation water can be supplied.

- Moist cloudy summers make the oil from olives too acidic to use, and too much rain encourages mildew on vines and makes the wine watery.

- Frosts in May are disastrous for vines and frosts also reduce the commercial use of olives (although frost does not kill the olive trees).

- Hail can also be disastrous for vines.

Solutions

- Irrigation has been extended into areas such as southern Spain to increase the volume of crop production. Dams have been built across rivers, the water from winter rainfall is stored in reservoirs and the water is used for irrigation;

continues

continues

pipes carry water to individual farms where sprinklers are used to water soft fruit and citrus fruit, and also vegetables such as cauliflowers, peas and beans.

■ New crops such as rice and cotton can now be grown in irrigated areas around the Mediterranean; these have replaced the traditional crops of olives and vines.

activities

1   State the total rainfall.   [2]

2   State the difference (range) in temperature.   [2]

3   (a) Calculate the average summer temperature from May to September.   [2]

   (b) How does this temperature suit the growth of olives and grapes?   [3]

4   (a) State and describe the minimum temperature.   [1]

   (b) How does this temperature suit the citrus fruit crop?   [2]

5   (a) State the total rainfall in late winter and early spring (from November to May), and compare this to the total for the rest of the year.   [3]

   (b) How does this amount of rainfall suit the growth of olive trees?   [2]

6   (a) Describe the main negative impact of this climate.   [3]

   (b) What crops can be grown if irrigation is used to change the impact of the climate in the Mediterranean area?   [2]

HIGHER TIER

7   Describe TWO ways in which climate suits the type of farming in this area of the Mediterranean.   [6]

8   Carry out a search on the Internet and present information using ICT on 'How the climate influences the type of farming' as practised in a different area of the Mediterranean.

ROLE OF TECHNOLOGY IN MODERATING THE IMPACT OF CLIMATE ON FARMING

Examples of technology

Figure 29
An irrigated area (the plain of Valencia in Spain).

Irrigation is the method of technology used in areas where the climate is too dry for farming. Irrigation is the artificial addition of moisture to the land. Spain is one of the most irrigated countries in Europe with over 3 million hectares of land served by irrigation schemes. In the mountains above the narrow coastal plains in the south, dams have been built to create reservoirs storing water. These dams hold back the winter rainfall and snowmelt. The water is taken from the reservoirs during the hot dry summers and then channelled to small intensively cultivated fields, where oranges and lemons can be grown instead of vines and olives.

Figure 30
Plastic greenhouses in the Almeria area of Spain – the 'Costa Del Polythene'.

This area of Spain has the world's largest concentration of plastic covering the land; the plastic is stretched out over 3 metre high poles to create a hothouse environment which suits the growth of tomatoes, beans and peppers, etc. The polythene sheets are used to trap the heat from the sun as incoming shortwave radiation can penetrate the sheet but the outgoing longwave radiation is trapped by it. These crops can be harvested twice a year when they are out of season in the rest of Europe.

In Co. Armagh mushrooms are grown on a large-scale commercial basis in dark polythene tunnels. Mushrooms are a type of fungus and do not require photosynthesis. This means light is only needed to manage the crop, e.g. to add water, to check for pests and disease, and to pick the mushroom crop. The polytunnels enable mushrooms to grow all year round when the outside air temperature falls too low. Heaters and coolers in the polytunnels keep the temperatures to at least 23°C to start the mushroom spores in the compost. The moisture content of the air is controlled using air vents, and humidity is kept high (over 90%) so that conditions are warm and moist to suit commercial production of mushrooms.

Figure 31
The polytunnels needed to grow mushrooms in Co. Armagh.

All these methods of technology are expensive to construct and to maintain, e.g. the plastic is easily damaged by strong winds and must be replaced every 2–3 years. This means the types of crops grown using technology to change the climate are those that will command a high price when sold by the farmers.

UNIT THREE
Impact of Human Activities upon the Atmosphere and the Environment

GLOBAL WARMING

The 1990s was the hottest decade on record and the year 2000 was the wettest year on record in England and Wales since 1872; it brought floods to many areas, destroying people's possessions, and ruining their homes and livelihoods. **Global warming** became front page news as the Deputy Prime Minister of the UK said that the country's power lines, drainage systems and flood defences could no longer cope with the more extreme weather in Britain.

Global warming means the increased heating of the atmosphere caused by human activities. World temperatures are estimated to have risen by 0.5°C in the 20th century and could rise by up to 5.8°C by the end of the 21st century. A 1°C rise in world temperatures could mean significant melting of the polar ice caps, worldwide rises in sea level and serious damage to existing ecosystems.

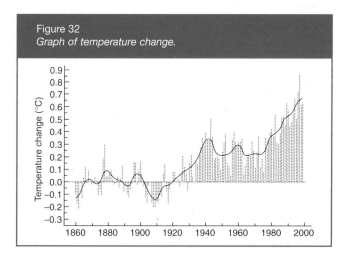
Figure 32
Graph of temperature change.

Describe how temperatures have increased over time (Fig. 32) [4]

The mechanism that creates global warming is the greenhouse effect.

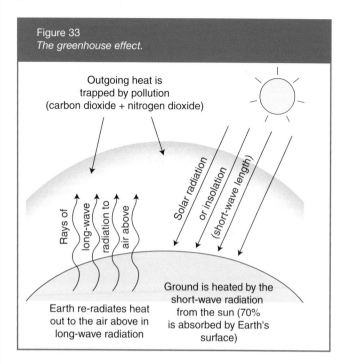

Figure 33
The greenhouse effect.

Outgoing heat is trapped by pollution (carbon dioxide + nitrogen dioxide)

Rays of long-wave radiation to air above

Solar radiation or insolation (short-wave length)

Earth re-radiates heat out to the air above in long-wave radiation

Ground is heated by the short-wave radiation from the sun (70% is absorbed by Earth's surface)

activities

Explain how the greenhouse effect contributes to global warming. [6]

Causes of global warming

Global warming is caused by greenhouse gases which trap heat in the atmosphere; the main greenhouse gases are carbon dioxide, nitrogen dioxide and methane. Carbon dioxide is responsible for 50% of global warming. It is feared that as rainfall belts shift, areas now covered by tropical rainforest could change into grassland or even desert. These would accelerate the rates of warming as fewer trees means an increase in the amount of carbon dioxide in the atmosphere. Methane production from large numbers of cattle is increasing very fast and so is nitrogen dioxide given off by fertilisers.

There are two human activities which are major sources of greenhouse gases:

■ Burning fossil fuels

In the last 200 years the need for more energy has grown as industrial development, population growth and prosperity have all increased. Most of this energy has come from burning fossil fuels. When coal, oil or gas is burned in power stations

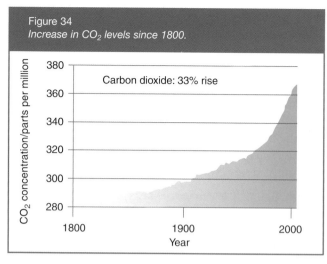

Figure 34
Increase in CO$_2$ levels since 1800.

Carbon dioxide: 33% rise

CO$_2$ concentration/parts per million

Year

Figure 35
Met Office press release, 19 December 2000.

There have now been 22 years in a row with above average global temperatures. The ten warmest years on record since 1860 have all occurred since 1983, and eight of the top ten have occurred since 1990. When we look at the patterns of change, we increasingly believe that a large part of the recent warming is due to fossil fuel burning.

Source: Met Office

to generate electricity, gases such as carbon dioxide are emitted into the atmosphere and form a blanket of pollution.

■ Vehicles

During the 20th century more and more vehicles were using the roads, a trend that is continuing into the 21st century. The exhausts of cars and lorries emit polluting gases such as nitrogen dioxide which add to the pollution in the atmosphere.

Figure 36
A cartoonist's view of global warming.

Figure 37
Some likely effects of global warming, as predicted by the IPCC.

EUROPE
- Melting of many Alpine glaciers and areas of permafrost
- Increased river flooding
- Droughts more common in southern Europe
- More deaths from heat stroke in cities during the summer months
- Spread of malaria

AFRICA
- Decrease in crop and plant yields
- Spread of desert margins
- Increased risk of flooding and coastal erosion in Senegal, Gambia and Egypt

AUSTRALIA
- More intense rainstorms and cyclones
- Droughts become more common

ASIA
- Decrease in agricultural output in South Asia
- More powerful storms during the monsoon season
- The swamping of mangrove forests that protect coasts and river banks, especially in Bangladesh
- The melting of glaciers in the Himalayas could trigger massive floods, affecting over half a billion people

THE AMERICAS
- An increase in crop output in some parts of North America, but a decrease due to drought in the Canadian Prairies and Great Plains
- An increased risk from storm surges, particularly in Florida and on the north-eastern seaboard

activities

1 Annotate a world map to show the impact of global warming using the information from the table in Fig. 37.

2 (a) Use the map of the British Isles (Fig. 38) to sort the impacts of global warming into two groups – Positive and Negative. [6]

HIGHER TIER

(b) Sort the impacts into two groups – those that affect people (social effects) and those that affect the environment. [4]

Figure 38
The possible advantages and disadvantages of global warming in the UK.

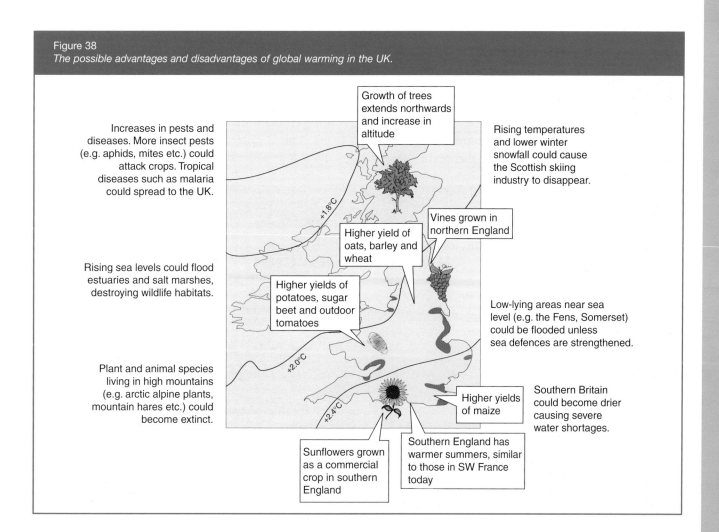

Increases in pests and diseases. More insect pests (e.g. aphids, mites etc.) could attack crops. Tropical diseases such as malaria could spread to the UK.

Rising sea levels could flood estuaries and salt marshes, destroying wildlife habitats.

Plant and animal species living in high mountains (e.g. arctic alpine plants, mountain hares etc.) could become extinct.

Growth of trees extends northwards and increase in altitude

Rising temperatures and lower winter snowfall could cause the Scottish skiing industry to disappear.

Vines grown in northern England

Higher yield of oats, barley and wheat

Higher yields of potatoes, sugar beet and outdoor tomatoes

+1.8°C

+2.0°C

+2.4°C

Low-lying areas near sea level (e.g. the Fens, Somerset) could be flooded unless sea defences are strengthened.

Higher yields of maize

Southern Britain could become drier causing severe water shortages.

Sunflowers grown as a commercial crop in southern England

Southern England has warmer summers, similar to those in SW France today

Impact or effects of global warming

Sea level changes

Sea levels could rise by up to 1.5 metres because the sea expands as it is heated; also ice caps at the poles and glaciers could melt leading to sea levels rising by up to 5 metres. Low-lying areas such as Bangladesh and the Netherlands could suffer from floods, and small islands such as the Maldives could disappear.

Climate changes

Patterns of world precipitation could change and rainfall become more unreliable so that some places will be wetter and others drier. Certain areas of the world could become drier with severe drought and heat waves. The number of people living in areas where water is scarce will increase from 1.7 billion to 5.4 billion within 25 years. Other places will experience strong winds and storms, for example more severe hurricanes on the crowded coastlines and islands in the eastern USA and Caribbean.

Tourism

Places such as the Alps could have less snow due to the rising temperatures, so reducing the conditions for skiing; this could have negative effects on employment, economic prospects and income for this region.

Agriculture

The types of crops grown and their yields will be affected by increased temperatures; more extreme climatic conditions may affect how the land is farmed, e.g. the need for expensive irrigation schemes to cope with drought could force farmers out of business. Countries in southern Europe could be at greater risk of water shortage and the quality of the soil could fall, severely affecting agriculture. Farmland could turn to desert in East Africa.

Health

The increasingly hot temperatures in northern Europe could put people at greater risk from heatstroke in cities, bringing a rise in deaths. Hot wet conditions can spread pests and diseases, e.g. more insect pests could attack crops. Mosquitoes which spread malaria could survive further from the equator. The West Nile virus arrived in the USA in 1999.

Economy

Damage caused by extreme climatic conditions will put financial pressure on governments, industries and individual property owners. More money may need to be spent on expensive flood protection schemes to protect coastal cities, defend low-lying coastal land, and safeguard ports and harbours. Property worth £220 billion and farmland worth £7 billion is at risk in the UK.

Solutions

Measures to reduce global warming

Examples:

- Limit greenhouse gases, e.g. countries could be given an annual limit to their production of carbon dioxide; if this limit is exceeded they may

---

Figure 39
Newspaper article on Bio diesel.

Cheaper fuel from recycled chip fat

CHEAP fuel in return for a bucket of old chip pan oil is being offered to motorists in Northern Ireland.

Special buckets are being given out by a Tyrone fuel company to collect used cooking oil to be recycled into "bio diesel".

A full bucket of fat collected from the frying pan or chip pan earns – under a special promotion – seven pence per litre off a fill-up at the pumps.

O'Neill's Fuels in Coalisland is believed to be the first company in Ireland to produce and sell bio diesel on a commercial basis – and it costs a lot less than ordinary diesel to buy.

The company has been developing the product for the past two years and has just put it on the market – and motorists can't get enough.

Not only does it cost a lot less, but it appears to provide the odd extra mile to the litre.

Director Tracy O'Neill said she had put it on sale at their own garage but hoped to eventually supply every filling station in Northern Ireland.

Bio diesel is suitable for all cars and is around 20p cheaper than the ordinary stuff.

"The benefits of this fuel are numerous," she said.

"It's low-sulphur fuel, therefore there are lower emissions into the atmosphere. It's also cheaper."

"Most people just dump their used cooking oil so this is a good chance for them to play their part in helping the environment."

have to pay a fine or buy part of the 'unused' pollution limit from a country below its annual pollution limit;

- Promote renewable forms of energy – one problem is that fossil fuels are relatively cheap compared to solar, tidal or wave power;

- Afforestation – plant more trees to act as 'sinks' of carbon dioxide;

- Increase fuel efficiency of vehicles, e.g. by promoting biodiesel and hydrogen fuel cells;

- Encourage greater use of public transport;

- Develop more organic farming using less nitrogen fertilisers.

Management response – the need for international cooperation

The Kyoto Agreement of 1997

Many countries signed up to reduce emissions of greenhouse gases by 5.2% and are taking steps to introduce measures to slow down global warming.

- The USA signed up to the agreement but is still reluctant to make any cuts because it is worried that cuts will damage its economy. This is despite the fact that the USA produced 21% of global carbon dioxide in 1996 but had only 4% of the world's population. By 2002 the USA was the world's largest producer of carbon dioxide both in total and per person; it produced over 15 tonnes of carbon dioxide per person per year in 2002.

- Some LEDCs are trying to develop their economy and feel they need to increase their use of energy to create new jobs and improve their standards of living.

- It was suggested that compensation be paid to countries that lose money if less of their oil is used and payments made to LEDCs to help them adapt to climatic change.

Figure 40
Greenhouse gas emission targets set at Kyoto, 1997.

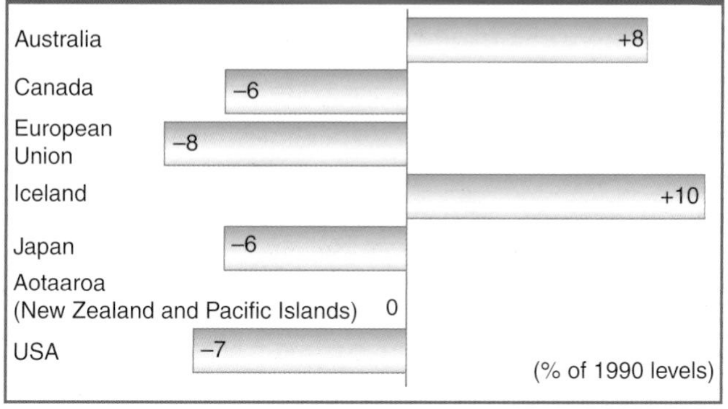

Australia	+8
Canada	−6
European Union	−8
Iceland	+10
Japan	−6
Aotaaroa (New Zealand and Pacific Islands)	0
USA	−7
	(% of 1990 levels)

- Major carbon dioxide polluters (e.g. Japan) are able to buy carbon credits from less-polluting countries (e.g. Australia).

The World Summit for Sustainable Development (WSSD) in Johannesburg, 2002

This conference established principles for sustainable development.

A **sustainable** solution to global warming means: 'Keeping greenhouse gases at levels which will not dangerously upset the global climate and yet will allow economic development to continue in both LEDCs and MEDCs'.

Principles established at the conference:

1 The polluter should pay, i.e. countries emitting pollution should pay a fine to control emissions.

2 Forests should be used as carbon 'sinks' – planting trees to absorb carbon dioxide.

3 Emissions of greenhouse gases should be reduced to 1990 levels.

4 LEDCs should be helped to become more energy efficient.

5 Invest in renewable energy resources.

6 Prepare to deal with the problems caused by global warming.

web link & extra resources

www.ctrcaltesc.org

www.iisd.org/rio+5/agenda/climate.htm

www.ciel.org

activities

1 Describe fully two sustainable solutions to the problem of global warming. [6]

2 Explain why (a) the USA and (b) the LEDCs have important parts to play in reducing global warming. [6]

HIGHER TIER

3 To what extent are each of the principles established at WSSD sustainable? [6]

4 Explain why it is so difficult to reach sustainable solutions to global warming. [6]

UNIT ONE
Crustal Movements and the Impact on People and the Environment (Management Issues)

CAUSES OF PLATE MOVEMENT

The Earth is made up of a series of layers.

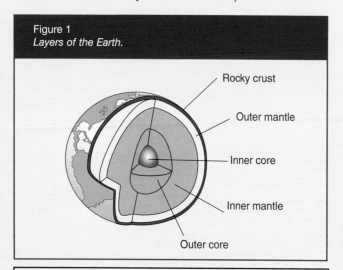

Figure 1
Layers of the Earth.

- Rocky crust
- Outer mantle
- Inner core
- Inner mantle
- Outer core

Did you know you are moving even though you think you are sitting still on your chair in a classroom? The plate you are on is moving at 70 mm per year, the Earth is revolving round the Sun at 19 miles a second. Our solar system is revolving around the centre of the galaxy at 185 miles per second and even our galaxy is moving through the universe at 375 miles per second!

Plate tectonics theory proposes that the Earth's crust is split into sections called **plates**. These plates are constantly moving around on top of the mantle, at an average speed of 70 mm per year. The places where plates meet, called plate boundaries, are related to seismic (**earthquake**) and volcanic activity.

The consequences of such movements of landmasses and having new oceans and seas opened are far reaching. It has influenced climate change and even the spread of plants and animals.

What causes plate movement?

In a more detailed cross-section of the Earth (Fig. 2) it can be seen that the crust on which we live acts as though it is floating on a layer of molten material, called the mantle. Inside the Earth there are convection currents within the mantle, moving heated molten material upwards from the core, up towards the crust. Here it cools and sinks back down

to the core, so that the cycle can start again. It is these convection currents that cause the crust above them to move.

Where currents descend (go downwards) they drag crust into the mantle, creating a destructive plate boundary. Here crust is destroyed as it descends down into the hot mantle, where it melts. Where currents ascend (rise up) they pull the crust apart, creating a constructive plate boundary. Molten material from the mantle rises to plug the gap in the crust, creating new crust.

It was these same convection currents that first broke up the crust, creating plates. The theory that explains this process and the related landforms and hazards is **plate tectonics theory**.

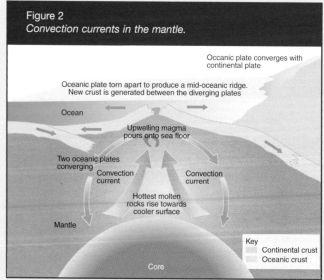

Figure 2
Convection currents in the mantle.

- Occanic plate converges with continental plate
- Oceanic plate torn apart to produce a mid-oceanic ridge. New crust is generated between the diverging plates
- Ocean
- Upwelling magma pours onto sea floor
- Two oceanic plates converging
- Convection current
- Convection current
- Hottest molten rocks rise towards cooler surface
- Mantle
- Core

Key
- Continental crust
- Oceanic crust

Types of plate boundary

There are three main types of plate boundary.

- Constructive – when plates are pushed apart, so they move away from one another, new crust is created.

- Destructive – when plates crash into one another and crust is destroyed

- Conservative – when plates slide past each other. Crust is neither created nor destroyed.

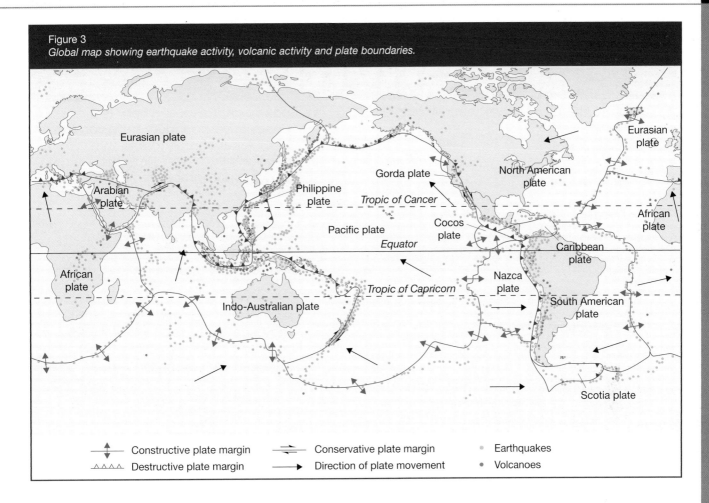

Figure 3
Global map showing earthquake activity, volcanic activity and plate boundaries.

Legend:
- Constructive plate margin
- Destructive plate margin
- Conservative plate margin
- Direction of plate movement
- Earthquakes
- Volcanoes

Features of plate boundaries – earthquakes and volcanoes

Earthquakes are the shaking of the ground surface caused by a sudden movement of the Earth's crust. Earthquakes can happen anywhere but, as Fig. 3 shows, they are mostly found along three main belts:

- Encircling the Pacific Ocean;

- Along the coast of the Mediterranean Sea and through southern Asia, towards the Pacific Ocean;

- Through the middle of the Atlantic, Southern and Indian oceans.

Figure 9 also shows the location of some of the most recent earthquakes. Remember, earthquakes are very frequent events. There are more than 150,000 earthquakes recorded globally each year. The world's main fold mountains have been formed by the buckling of the Earth's surface during crustal movements felt as earthquakes.

Volcanoes are mountains, often cone-shaped, formed by surface eruptions of magma from inside the Earth. During eruptions, lava, ash, rock and gases may be ejected from the **volcano**. Figure 3 shows the location of areas that experience volcanic activity. They also form three main belts:

- Around the edge of the Pacific Ocean – known as the 'Ring of Fire' because there are so many active volcanoes;

- Through the Mediterranean Sea, and down the east coast of Africa;

- Down the middle of the Atlantic Ocean.

Volcanoes are also found in isolated clusters, such as the Hawaiian Islands in the middle of the Pacific Ocean, and Réunion in the Indian Ocean.

When we look at the locations of areas that have both earthquakes and volcanoes (zones of activity) we see a definite pattern which corresponds to plate boundaries.

activities

1 (a) State the meaning of the following terms:

   Plate   [2]

   Earthquake   [2]

   Volcano   [2]

   (b) Draw a diagram to help explain why plates move. Ensure it contains the following labels:

   Destructive boundary
   Constructive boundary
   Convection current
   Mantle
   Crust   [6]

2 (a) Using the map shown in Fig. 3 on the previous page, state three areas in the world where we can find both volcanoes and earthquakes.   [6]

   (b) Copy and complete, by choosing the correct ending, the following statement describing the relationship between earthquakes, volcanoes and plate boundaries.   [1]

   'Most earthquakes and volcanoes occur **in the middle of plates/at plate boundaries**.'

3 (a) Look back at Fig. 3. There is a constructive boundary between the North American plate and the Eurasian plate. Complete the following paragraph about this boundary, filling in the correct word in the blanks.

   At a constructive boundary tectonic plates are moving _____. The plates are being pulled by _____ currents in the mantle below. As they move apart the plates get further away from one another. One example is seen in the middle of the Atlantic Ocean, where the North American plate is moving away from the _____ plate. This means in the UK we are 3 cm further away from _____ than we were last year.   [4]

   (b) Write a similar paragraph, this time for a destructive boundary.   [6]

HIGHER TIER

4 Describe and explain the global distribution of earthquakes and volcanoes. If you have access to a computer, include diagrams and information from the weblinks below. [12]

web link & extra resources

www.enchantedlearning.com/subjects/dinosaurs/glossary/pangaea.shtml

www.geography.learnontheinternet.co.uk/topics/PLATEte.html

www.historyoftheuniverse.com/pangagea.html

www.georesources.co.uk/tectonicg.htm – the home page here leads you into this very good website which contains more detail on this section of the course.

Features and characteristics of each plate boundary

The features and hazards of each plate boundary depend on the type of boundary (constructive, destructive or conservative) and the type of crust involved (oceanic or continental).

The table on page 33 summarises the characteristics of the two different types of crust.

Constructive boundary

One constructive boundary is found in the middle of the Atlantic Ocean. Here the Eurasian and North American plates are being pulled apart, moving away from one another. This means the Atlantic Ocean is getting wider, by about 3 cm a year. This movement causes regular, but weak earthquake activity. Magma wells up from the mantle to plug the gap, so there is

Oceanic crust	Very dense (heavy). Mean density is 3000 kg m^{-3}.	Thin: 5–10 km	Can sink into the mantle	Easily destroyed	Young crustal material
Continental crust	Less dense (light). Mean density is 2700 kg m^{-3}.	Thick: 30–70 km	Does not sink easily into the mantle	Hard to destroy	Old crustal material

often frequent, gentle volcanic activity even here under the ocean. This rising of material causes the crust to rise slightly at either side of the plate boundary creating an oceanic ridge. The volcanoes found along the ridge are called smokers. This chain of volcanic mountains is the longest in the world and means that the middle of the Atlantic is relatively shallow. The hardened lava erupted from volcanoes forms new crust.

Constructive boundaries are also found where continental crust is splitting apart.

Rift valleys are seen at constructive margins on continental crust. One example is the Great Rift Valley in Eastern Africa.

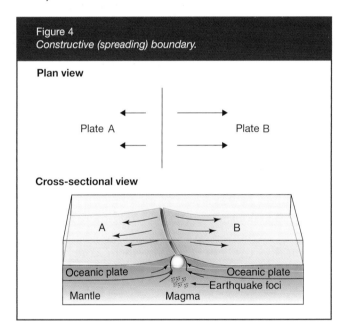

Figure 4
Constructive (spreading) boundary.

Plan view

Plate A   Plate B

Cross-sectional view

A   B
Oceanic plate   Oceanic plate
Earthquake foci
Mantle   Magma

Destructive boundary

Destructive boundaries have a zone of subduction, here crustal material is being pulled into the mantle, where it melts and is destroyed.

There are three types of this type of boundary:

1 Oceanic crust crashing into continental crust;

2 Oceanic crust crashing into oceanic crust;

3 Continental crust crashing into continental crust.

An example of Type 1, where oceanic crust is crashing into continental crust, can be found on the western

coast of South America. Here the Nazca plate, made of oceanic crust, is disappearing below the South American plate. At the plate boundary, the heavy oceanic crust is being pushed downwards into the mantle. It is dense (heavy), and so it falls below its normal level as it sinks into the mantle, creating a deep ocean trench called the Peru–Chile Trench. This linear trench follows the line of the western coast, and is very deep in places. It is up to 8050 metres deep!

The movement of the Nazca plate against the South American plate is not smooth, because of the friction between the rough surfaces. The plates may become stuck for years, until the pressure for the plates to move is greater than the friction preventing them from moving. The pressure is released suddenly and the two plates will jolt many centimetres at once. This sudden movement is felt on the Earth's surface as an earthquake. The further inland the earthquake occurs, the deeper its focus will be (the point of origin of an earthquake below the Earth's surface) and the weaker the shockwaves are when they reach the surface – meaning a less destructive earthquake.

As the Nazca plate is subducted (disappears down) into the mantle, it begins to melt due to intense heat.

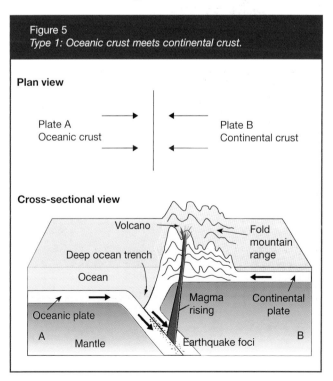

Figure 5
Type 1: Oceanic crust meets continental crust.

Plan view

Plate A
Oceanic crust   Plate B
Continental crust

Cross-sectional view

Volcano   Fold mountain range
Deep ocean trench
Ocean
Oceanic plate   Magma rising   Continental plate
A   Mantle   Earthquake foci   B

Because it was once oceanic crust, it is saturated with water. The magma created is therefore chemically different from any naturally found in the mantle – it is full of gas bubbles, created by the evaporating water. Together these mean that this melted oceanic crust is less dense (lighter) than the surrounding mantle, so it rises upwards as an explosive type of magma. If it breaks through the surface, it creates a volcano (e.g. Cotopaxi and Chimborazo in Ecuador). The continental crust that makes up the South American plate is not dense like oceanic crust, so it is not subducted easily. Instead it folds and buckles upwards to create a linear fold mountain range, such as the Andes.

> **Note**
> Types 2 and 3 are recommended for Higher Tier only.

Type 2, where oceanic crust and oceanic crust meet, has many similar features to Type 1. An arc of volcanic islands and deep ocean trench are located at this boundary. In fact the oceanic trench can be even deeper than that seen in Type 1; for example, the deepest part of the Mariana Ocean Trench, in the Pacific, is 11022 m. This could easily swallow Mt Everest, the tallest mountain on the surface, which is 8848 m high. There are no fold mountains as there is no continental crust to buckle upwards.

Where two continental plates meet (Type 3), the crust of both plates buckles and folds upwards. The two sets of fold mountains overthrust one another, creating a large range of high mountains. There is little material melting, and that which does cannot make it through the high mountains to create a volcano. Instead the

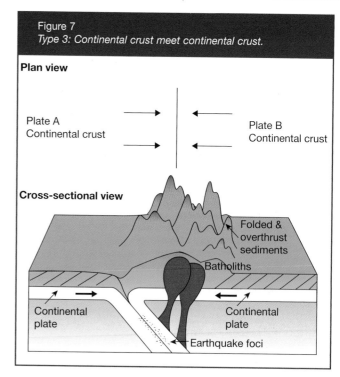

Figure 7
Type 3: Continental crust meet continental crust.

magma forms large intrusions into the mountain range, called batholiths. The magma cools slowly to form granite cores to the mountains. Good examples of mountains formed at this type of boundary include the Himalayas and the Zagros ranges.

Conservative boundary

At conservative boundaries, such as the San Andreas fault line in California, two plates try to slide past one another. When friction causes the two plates to stick, pressure to move builds up. This pressure is eventually released as an earthquake when the plates move suddenly. As crust is neither created nor destroyed at conservative boundaries, there are no volcanic eruptions. Over 20 million people live along the San Andreas fault. The movement along this conservative boundary can be seen in offset streams and orchards.

EARTHQUAKES

Earthquakes are a natural hazard. Major earthquakes can release the same amount of energy as a large nuclear explosion. The destructive power of an earthquake is measured by the amount of energy it releases. Earthquakes are recorded using seismographs.

The paper tracing of the seismic waves (seismogram) is then converted to a level on the Richter scale. The Richter scale ranges from 0 to 9, with each point on the scale representing an earthquake ten times the magnitude and which releases 30 times more energy than the point before. Figure 9 shows the whole Richter scale.

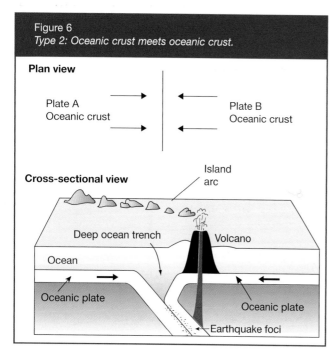

Figure 6
Type 2: Oceanic crust meets oceanic crust.

activities

1   State five differences between oceanic and continental crust.   [10]

2   (a) Name two places a constructive boundary exists, and for each name a landform found there.   [6]

   (b) Explain why volcanic activity is commonly found at plate boundaries.   [4]

3   (a) Make a labelled copy of a destructive plate boundary where oceanic and continental crust meet.   [7]

   (b) Choose one of the features you have named and state fully why it is found at this plate boundary.   [4]

4   Complete this summary diagram relating to plate boundaries. One has been filled in for you.   [10]

HIGHER TIER

5   State which of the following features/hazards named below are not found at a destructive boundary that only involves oceanic crust. Explain your answer.

   volcanoes, earthquakes, ocean trench, fold mountains.   [8]

Plate boundary	Types of crust involved	Global example	Earthquake activity	Volcanic activity	Other features
Constructive	Oceanic & oceanic	Mid-Atlantic Ridge	Gentle and shallow	Gentle and regular	Ocean ridge
Constructive					
Destructive					
Destructive					
Destructive					
Constructive					

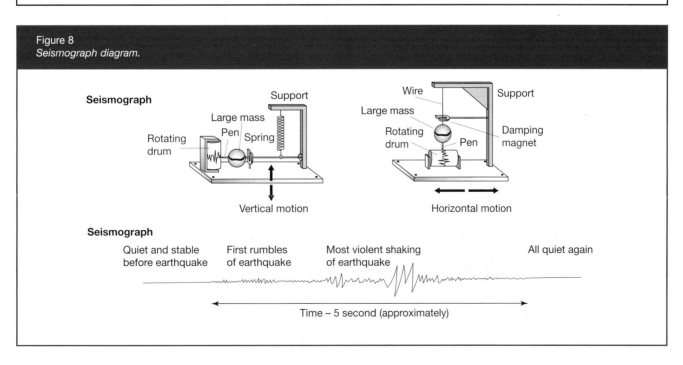

Figure 8
Seismograph diagram.

Seismograph

Support
Large mass
Pen   Spring
Rotating drum
Vertical motion

Wire
Support
Large mass
Rotating drum
Pen
Damping magnet
Horizontal motion

Seismograph

Quiet and stable before earthquake   First rumbles of earthquake   Most violent shaking of earthquake   All quiet again

Time – 5 second (approximately)

Figure 9
The Richter scale measures the size of the seismic waves during an earthquake.

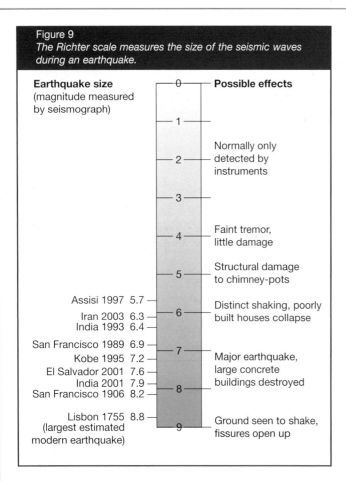

Earthquake size
(magnitude measured
by seismograph)

Possible effects

0
1
2 — Normally only detected by instruments
3
4 — Faint tremor, little damage
5 — Structural damage to chimney-pots
Assisi 1997 5.7
6 — Distinct shaking, poorly built houses collapse
Iran 2003 6.3
India 1993 6.4
San Francisco 1989 6.9
7 — Major earthquake, large concrete buildings destroyed
Kobe 1995 7.2
El Salvador 2001 7.6
India 2001 7.9
8
San Francisco 1906 8.2
Lisbon 1755 8.8
(largest estimated modern earthquake)
9 — Ground seen to shake, fissures open up

The main features of an earthquake are shown in Fig. 10. The place where an earthquake starts is termed its focus. Shock waves spread from this point, like ripples on a pool after a stone is dropped in. The most deaths and maximum destruction are normally seen right at the epicentre, the point directly above the focus, where shock waves are first felt on the surface. The amount of damage decreases as distance increases from the epicentre. This is described as a negative relationship.

Figure 10
Features of an earthquake.

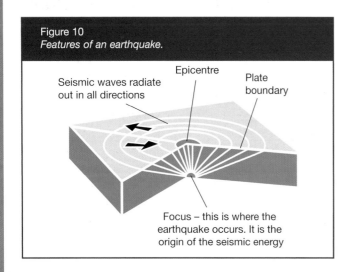

Seismic waves radiate out in all directions

Epicentre

Plate boundary

Focus – this is where the earthquake occurs. It is the origin of the seismic energy

activities

1 (a) State the use of a seismograph.

(b) Explain how a seismograph works. [2]

2 Give definitions of the following earthquake terms:

• Richter scale

• Epicentre

• Focus

• Shock wave. [4]

3 (a) Read the following extract. How strong do you think the earthquake was on the Richter scale? Justify your answer.

'The earthquake woke me. I could feel the ground shaking and one of my wardrobes fell over. I crawled under my bed for protection. The tremors lasted for a minute or so. When they were over, our family gathered outside the house to see what had happened. Some of our plates were broken inside the cupboards, and our chimney was damaged.' [3]

(b) Write your own extract to illustrate what it might be like to experience an earthquake that is at a different level on the Richter scale. [6]

HIGHER TIER

4 (a) State fully why there is more damage close to the epicentre of an earthquake. [5]

(b) Why do you think that not all earthquakes are reported on the news? [4]

Earthquakes can have a wide variety of impacts on people and the environment. These are outlined in the case studies which follow.

CASE STUDY

AN EARTHQUAKE EVENT IN A MEDC - KOBE, JAPAN

The Kobe earthquake occurred at 5.46 am on 17 January 1995. During the quake the ground moved up to 1.2 metres vertically and 2.1 metres horizontally. It measured 7.2 on the Richter scale and lasted only 20 seconds. The damage was so great because it was a very shallow earthquake, the focus was near to the surface. The epicentre was 20 km SW of downtown Kobe, making it very close to the second most densely populated area of Japan.

Causes

Japan is next to a destructive plate boundary, where three plates meet (see Fig. 11). Just off the Kobe coastline a dense oceanic plate, the Philippine plate, is being subducted under the lighter continental Eurasian plate at a rate of approximately 10 cm per year. The movement of one plate against the other has lead to many earthquakes in this region of Japan.

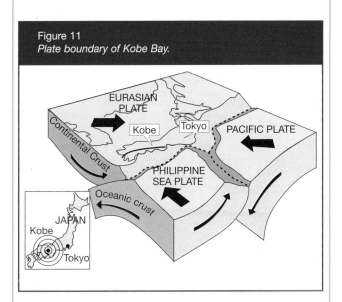

Figure 11
Plate boundary of Kobe Bay.

Short-term impacts

5500 people died and a further 40,000 were injured. Some died as buildings, roads and bridges collapsed, but more died in the fires that broke out all over the city.

continues

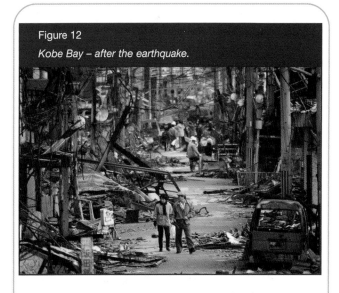

Figure 12
Kobe Bay – after the earthquake.

200,000 buildings collapsed. Many of these were the two-storey wooden houses built just after the Second World War. They only had thin walls, but a heavy roof made of tiles. During the quake the roofs tended to fall in, collapsing the walls as they did so. People on the upper floor often survived, but the elderly, who were sleeping on the ground floor, were not so lucky. Almost half of the 5500 people who died in the earthquake were the elderly who customarily sleep on the ground floor. The office blocks built in the 1960s tended to collapse in the middle. Only the most modern buildings suffered little damage, because they were designed to be earthquake proof.

Transport routes were disrupted. A 1 km stretch of the Hanshin Expressway collapsed and the track of the Shinkansen bullet train snapped in eight places. The roads became grid locked, delaying ambulances and fire engines.

At the main port there was significant damage. **80% of the quays where ships were moored were destroyed**. The wetter ground here suffered liquefaction, that is when shaking brings underground water to the surface, turning what was solid ground into mud. Industry in the area, including major firms such as Mitsubishi and Panasonic, was forced to close.

Electricity, gas and water supplies were disrupted. Many residents had only bottled drinking water, rather than mains water, for up to 2 weeks. More than 150 fires, started by the broken gas mains and sparks from cut-off electrical cables, raged for days in the city, destroying an estimated 7500 homes. These fires

continues

released acrid (foul smelling) smoke into the air, creating a smog so thick that it obscured the Rokko Mountains behind the city.

Around a quarter of a million people were made homeless. Some managed to gain shelter in emergency accommodation centres set up in larger public buildings like schools and town halls. Many others had to spend the night outdoors at a time when night temperatures drop to –2°C. Throughout there was a shortage of blankets, food and water.

Long-term impacts

It took until July before water, electricity, gas and telephone services were fully operational, and until August for rail services to get back to normal.

One year later the port of Kobe was 80% functional, but the Hanshin Expressway remained closed.

It took a long time to rebuild the houses and repair buildings. Even in January 1999 some people were still living in temporary accommodation.

Gaps in society exist because of deaths and also people unable to return to normal functions because of injury. Each year on the anniversary of the earthquake, the people of Kobe hold a special candlelit ceremony to remember those who died. For the survivors, life will never be the same again.

What was the management response?

The Japanese have always had to live with the threat of earthquakes. Once they attributed the earth tremors to the thrashing of a giant catfish called Namazu. Today, the Japanese have put their faith in science and research to help them predict earthquakes, engineer earthquake-proof buildings and co-ordinate the most efficient quake relief.

Before the earthquake – prediction and precautions

The Japanese are world leaders in designing buildings to withstand earthquakes, using springs or rubber pads to absorb the shock of the tremors. However, the Kobe earthquake was the worst to hit Japan since 1948. A section of the elevated Hanshin Expressway collapsed. Although the Expressway had been designed to withstand earthquakes, it had not been strengthened to current standards.

Minor tremors are felt in Japan nearly every day. As a result, millions of pounds are spent strengthening buildings by adding things like cross-beams that will spread any shock waves more evenly across a building or by using specially reinforced concrete in construction. Buildings are stronger than in California and some of the more modern buildings survived the earthquake. The Kansai International Airport and Akashi Bridge, which were newly opened in 1995, were both undamaged, presumably due to their high-tech construction.

Japan has a public education programme, producing pamphlets, broadcasts and lectures about earthquake survival. On the anniversary of the Great Quake of 1923, each 1 September, there is a public holiday to practise earthquake drills.

Tokyo Metropolitan Government

What to do if a big earthquake hits

The worst shake is over in about a minute, so keep calm and quickly do the following:

- Turn off all stoves and heaters. Put out fires that may break out. Do not become flustered by the sight of flames, and act quickly to put out the fire.

- Get under a table or desk to protect yourself.

- Open the door for an emergency exit. Door frames are liable to spring in a big quake and hold the door so tight they cannot be opened.

- Keep away from narrow alleys, concrete block walls and embankments, and take temporary refuge in an open area.

- For evacuation from department stores or theatres, do not panic and do as directed by the attendant in charge.

- When driving in the streets, move the car to the left and stop. Driving will be banned in restricted areas.

- Evacuate to a designated safety evacuation area when a big fire or other danger approaches.

- Walk to emergency evacuation area. Take the minimum of personal belongings.

- Do not be moved by rumours. Listen for the latest news over the radio.

Households are encouraged to keep earthquake kits, containing bottled water, rice, a battery powered radio, a fire extinguisher and blankets.

A 10-day supply of water is stored in underground cisterns and quakeproof warehouses.

continues

continues

Immediate and long-term action after the earthquake

An increased number of seismic instruments to record and measure earth movements were installed in the region.

In the late 1990s laws were passed that meant all buildings and transport structures had to be even more quakeproof:

- High-rise buildings had to have flexible steel frames;
- Small buildings had to have concrete frames with reinforced bars to absorb shock waves;
- Houses were to be built from fire-resistant materials, not just bricks and wood;
- New buildings had to be built on solid rock, not clay or landfill material.

CASE STUDY

AN EARTHQUAKE EVENT IN A LEDC - EL SALVADOR

El Salvador lies on the peninsula of land that joins the continents of North and South America. It is particularly prone to earthquakes and volcanic activity because it lies on the infamous 'Ring of Fire' – an arc of plate boundaries that encircle the Pacific Ocean.

It suffered a series of earthquakes in the early part of 2001. The strongest was on 13 January measuring 7.6 on the Richter scale. Lesser earthquakes were experienced on 13 February (measuring 6.6 on the Richter scale) and 20 February (measuring 3.6 on the Richter scale). The main shock, on 13 January, had a shallow focus – at a depth of 13 km. The epicentre was located some 30 km east of San Salvador in the region called La Paz. This quake was felt throughout El Salvador and in the neighbouring countries of Guatemala and Honduras. The quake only lasted for 10 seconds.

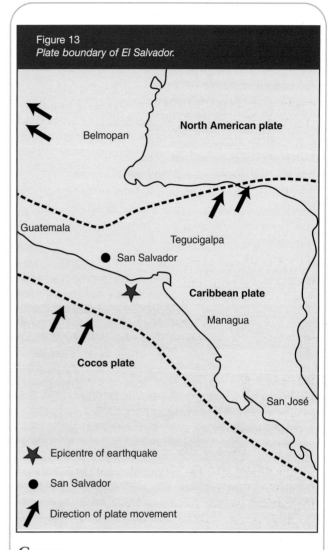

Figure 13
Plate boundary of El Salvador.

★ Epicentre of earthquake

● San Salvador

↗ Direction of plate movement

Causes

El Salvador is in a zone of activity. It is near a destructive plate boundary where the Caribbean and Cocos plates meet.

What were the short-term impacts?

At least **1000 people died and 3000 were injured**. The death toll was so high because the volunteer rescuers were only equipped with pickaxes and so did not have the heavy machinery that they needed to rescue people from under piles of rubble.

Thousands of buildings were destroyed, roads were blocked and powercuts were widespread. Official reports claim 13,935 houses were damaged and 38,699 houses were destroyed.

A centuries-old **church collapsed in Santa Ana** while people were worshipping inside.

continues

continues

Medical centres were overwhelmed by the thousands injured and were struggling to cope with the extra patients. Seventeen had to operate from temporary shelters.

There was a **mudslide** in San Salvador's middle class area. Tonnes of mud was dislodged above the Las Colinas suburb. The mud slid down the mountain in the first few seconds of the earthquake, burying as many as 400 homes. After 6 hours of digging through the debris of concrete blocks, mud and up-turned trees, only 20 bodies had been found out of the estimated 1200 feared buried under the landslide.

The Pan American Highway was blocked by three landslides, although they were cleared quickly.

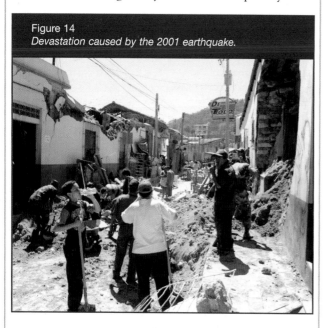

Figure 14
Devastation caused by the 2001 earthquake.

What were the long-term impacts?

There were at least 1 million people made homeless.

The cost of the earthquake is estimated at $2.8 billion – this is $1.3 billion more than the country's annual GNP.

Landslides around Lake Ilopango have altered drainage patterns that could cause more flooding here in the future.

Management response before the earthquake: predictions and precautions

Since the 1990s US AID has provided extensive training in El Salvador to increase their local capacity to prepare for and respond to natural disasters.

However, the country was relatively unprepared. El Salvador was still recovering from the damage caused by Hurricane Mitch in 1998. This hurricane had devastated the country.

There were regulations in place that forbid the building of housing in unstable areas that would be most at risk during an earthquake. According to some reports the developers at Las Colinas had been authorised to build despite zoning regulations in the area. The location was a desirable one because it was untroubled by guerrilla activity.

Also, El Salvador was just emerging from a period of civil unrest. The UN had international observers in the country in the early 1990s to ensure that the transition to democracy ran smoothly. Under these circumstances there was no long-term stability before the earthquakes for a government to implement effective strategies to minimise the damage that might be caused by such an event.

Management response after the earthquake: immediate and long term

Charities such as Oxfam provided clean water and distributed blankets, buckets, hygiene kits and plastic sheeting to help those who were made homeless, or who had been injured.

The United Nations appealed for extra funds to provide enough new housing, food and health care.

The medical centres struggled in the long term to care for patients, because they were still trying to treat injured from the first earthquake when the second and third quakes followed.

The Inter-American Development Bank approved a $20 million emergency loan for El Salvador. A $15 million portion of the loan will be used to provide families with basic housing kits and to fund the clean up and removal of debris. A $3.9 million portion will go toward the control of unstable hillsides in populated areas.

International response to the earthquakes El Salvador experienced in February was not as extensive as that given in January, because the earthquake in Gujarat, India had strained volunteer resources.

President Flores set up refugee camps.

continues

Contrasts in response between a MEDC (Japan) and a LEDC (El Salvador)

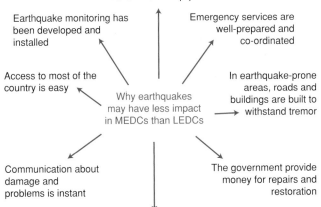

The emergency services have modern equipment

Earthquake monitoring has been developed and installed

Emergency services are well-prepared and co-ordinated

Access to most of the country is easy

Why earthquakes may have less impact in MEDCs than LEDCs

In earthquake-prone areas, roads and buildings are built to withstand tremor

Communication about damage and problems is instant

The government provide money for repairs and restoration

People are evacuated quickly and can avoid damage from secondary quakes

web link & extra resources

www.bbc.co.uk – search the news archives for details of the featured earthquakes or even more recent ones that you have heard about on the news. A good example would be to compare the earthquake in California to that of Bam in Iran – both happened in December 2003.

General sites on earthquakes – some have good interactive quizzes and maps:

www.quake.wr.usgs.gov/

www.creativeclassroom.starhub.com/disasters/EARTHQUAKEs01.asp

www.crustal.ucsb.edu/ics/understanding

www.gps.caltech.edu/seismo/seismo.page.html

www.civeng.carleton.ca/cgi-bin/quakes-101k – map showing earthquake locations over the preceding 30 days.

www.georesources.co.uk/kobehigh.htm – excellent detail on Kobe 1995.

http://ccs.cla.kobe-u.ac.jp/Asia/Visitor/Furm/report/yamamoto.html – read Manaby's real-life experience of the Kobe earthquake. You could then write your own imaginary experience of a similar earthquake.

www.redcross.org/services/disaster/keepsafe/readyearth.html

www.tremor.nmt.edu/prepare.html

activities

1 (a) Name an earthquake event you have studied. [1]

(b) State fully one effect it had on the environment. [3]

(c) State fully one effect it had on local people. [3]

2 It is possible to reduce the damage caused by earthquakes in MEDCs. For a MEDC you have studied, state fully two ways this has been attempted. [6]

3 The effects of earthquakes tend to be more severe in LEDCs. State fully two reasons why this is so. [6]

4 Go to the websites www.redcross.org/services/disaster/keepsafe/readyearth.html and tremor.nmt.edu/prepare.html. Read the instructions on how to prepare for an earthquake, and create your own information pamphlet. [10]

As a class you might be able to create your own earthquake survival kit.

HIGHER TIER

5 With reference to a named earthquake you have studied, describe the scale of the event and the effect it had on the local population and environment. [12]

6 Earthquakes in LEDCs often cause greater damage and loss of life than similar events in MEDCs. With reference to two contrasting areas you have studied, explain why this is so. [10]

EXTENSION

Using the BBC website, create a PowerPoint presentation comparing the earthquake in California to that of Bam in Iran – both happened in December 2003. [20]

UNIT TWO
Rivers and River Management

CHARACTERISTICS OF A DRAINAGE BASIN

Water is a critical resource. The water that is most useful to humans is fresh water, although this only makes up 2.8% of all the water on the planet, and only 0.1% is stored in rivers and lakes. The rest of the water on land is stored in ice sheets and glaciers, or in the soil and deeper down in the ground. The total amount of water on our planet never changes, in other words, none arrives from space, and none is lost to space. This is called a closed system. The water on Earth circulates between the sea, land and air (stores), being recycled in a natural process known as the **hydrological cycle** (water cycle).

On land water is stored on the surface as lakes and rivers. Each river is contained within its own **drainage basin** – the area of land drained by a river and its tributaries. The boundary of a drainage basin follows a ridge of high ground, known as the watershed. This and other features of the drainage basin are summarised in Fig. 15.

Simple open system of the drainage basin:

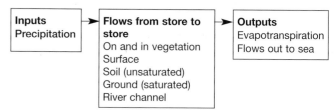

Water enters the drainage basin system as precipitation. This may be any form, such as snow or rain. Most drainage basins have some vegetation. The precipitation may be caught on the leaves of plants. This is called interception. It is generally greatest in summer. From the surface of the plant, the water may be evaporated back into the air, or flow down the stem of the plant to reach the ground. At this point the water has moved from the store in the

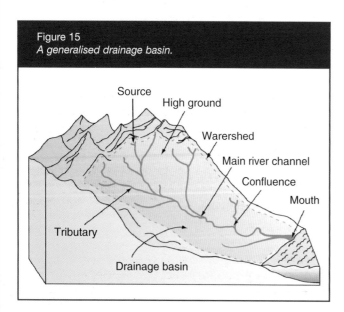

Figure 15
A generalised drainage basin.

The amount of water within a single drainage basin can vary, as it has inputs (from precipitation) and outputs (from evapotranspiration). So this is an open system.

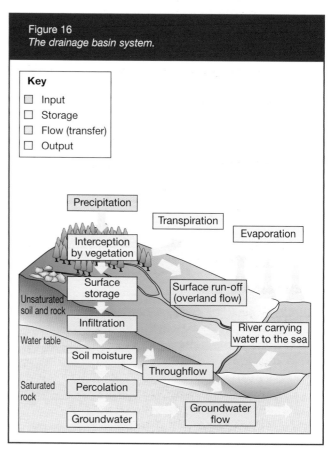

Figure 16
The drainage basin system.

vegetation to be part of the surface storage. If conditions are right, it will then seep into the soil. This process is called infiltration. Soil normally has small pockets of air called pores, which allow the water to get into it. Once in the soil, gravity will pull the water downwards and it will move down through the soil as throughflow, until it reaches the watertable, where all the pores in the soil, or rock, are already full of water, so it cannot move any further downwards. Instead it now flows laterally (sideways) into the nearest river as groundwater flow.

Any water that hits an impermeable surface, with no pores, such as tarmac, cannot infiltrate the soil below. It simply flows over the surface as surface run-off into the nearest river.

Although some precipitation can fall directly into the river, most water reaches a river by a combination of surface run-off, throughflow and groundwater flow. It takes water the longest time after falling to reach the river by groundwater flow, since it has had to flow through so many stores to get to the river channel.

RIVER CHANNEL CHANGES DOWNSTREAM

To investigate how a river can change downstream, it is possible to examine a local river like the Glendun River in Co. Antrim. The location of this river is shown on page 44. Any river may be divided off into three main sections called courses.

Measuring the Glendun River

Various **fluvial** characteristics are measured at regular points (every 1 km) along the Glendun River. This type of sampling is called systematic sampling and it allows the investigation of continuous changes as distance increases from the source of the river.

A group of pupils investigated this river. Figure 17 below shows what students measured.

How are these things measured?

Width

This is measured by placing one end of a measuring tape at one side of the river channel, then pulling it out to the other side of the channel. The distance is the width of the river.

activities

1 State the meaning of the following important river basin terms: drainage basin, watershed, source, tributary, confluence, river channel, mouth. [14]

2 (a) Name one input, two stores and two flows within the drainage basin system. Present your answer as a flow diagram. [7]

   (b) State fully two reasons why the total amount of water falling as precipitation never reaches the river channel. [6]

3 (a) The following statements describe conditions when there will be extra water flowing in a river channel. Put them in order, with the conditions that will create the greatest increase in flow first.

   • Rain on a summer's day.

   • Heavy storm rain in winter.

   • Snow in winter. [3]

   (b) Explain why you chose to order the statements as you did. [6]

HIGHER TIER

4 (a) Using evidence from the text, state fully why environmentalists are strongly against polluting rivers and lakes. [6]

   (b) Suggest the effect that building urban areas may have on the drainage basin cycle. [5]

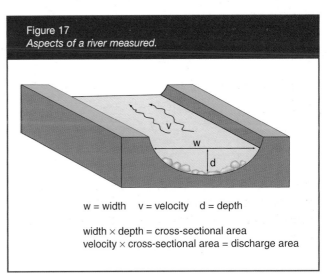

Figure 17
Aspects of a river measured.

w = width   v = velocity   d = depth

width × depth = cross-sectional area
velocity × cross-sectional area = discharge area

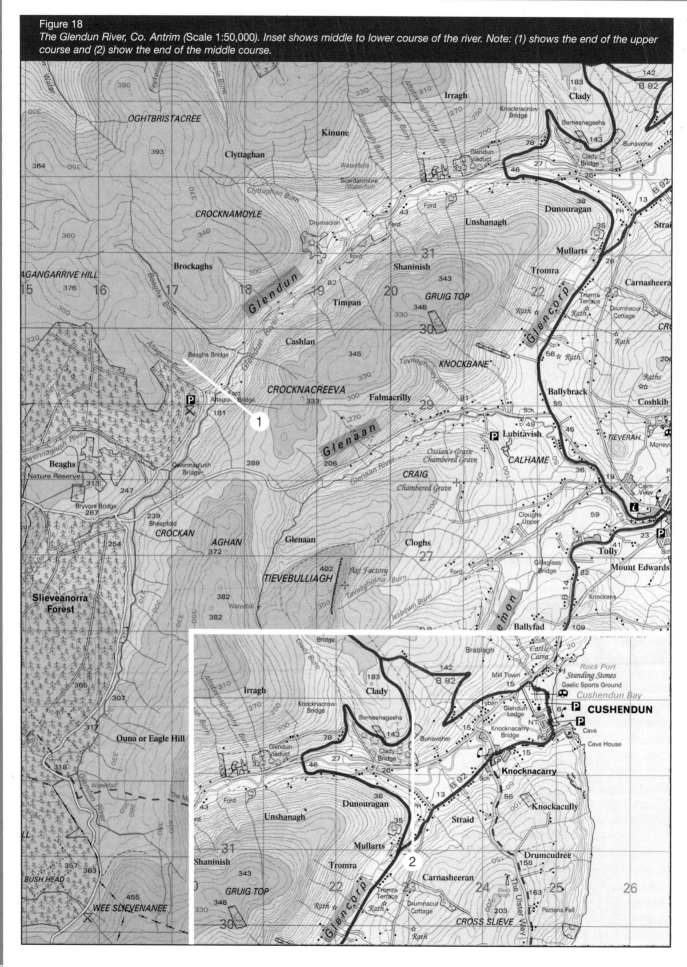

Figure 18
The Glendun River, Co. Antrim (Scale 1:50,000). Inset shows middle to lower course of the river. Note: (1) shows the end of the upper course and (2) show the end of the middle course.

Depth

This is completed using a metre stick. It is lowered into the water every 10 cm, and the distance from the top of the water to the river bed gives the depth of water. An average of all these readings is taken.

Discharge

Discharge is the amount of water passing any point in a river in a certain time, normally given as cubic metres of water per second (cumecs). It is calculated by multiplying the cross-sectional area of a river channel at a certain point by the speed (velocity) of the river at the same point.

The cross-sectional area is obtained by multiplying the width of the river by the average depth. The speed (velocity) of the river is recorded using a flow metre that when dipped into the river gives a digital reading of the speed of flow in metres per second.

Load

The load of a river is the material it is carrying, ranging from small sediment to large boulders. It is very hard to measure the size of the load in suspension, so instead, we can concentrate on the load lying on the channel bed – called bed load. This load is measured for size and roundedness. By measuring the longest axis of 15 random samples at each point an idea of the size of the load is obtained. Each stone is then given a rating for roundedness.

What were the results?

To help see the over all trends, here is a selection of results obtained from the Glendun. They represent the three courses of this river.

	Upper course (Station 1)	Middle course (Station 11)	Lower course (Station 16)
Width (m)	2.7	10.4	14.2
Depth (m)	0.14	0.33	0.46
Discharge (cumecs)	0.08	0.2	5.1
Load long axis (cm)	26	12	7
Load – roundedness	angular	sub angular	rounded

Are there any trends? How can they be explained?

To help understand and explain the results collected read the information about the processes a river carries out – **erosion**, transport and **deposition**.

Going downstream from source to mouth, it appears that the Glendun River gets wider. At Station 16 (16.5 km from the source) the river is just over five times the width it is at station 1, only 1.5 km from the source.

The river also appears to get deeper. At the station in the lower course the river is 32 cm deeper than it is in the upper course.

This can be explained by the fact that there is more lateral and vertical erosion occurring downstream from the source.

The enlarged river channel size downstream relates well to the pattern of increasing **discharge**. Because discharge is calculated by multiplying the cross-sectional area of the channel by the river's velocity, then it follows logically that as cross-sectional area increases so does the discharge. The river is receiving additional water from the tributaries that are entering it at regular intervals within the Glendun valley: these will also cause the discharge to be greater downstream. In the upper course very few tributaries have contributed to the flow. Finally, the velocity of the river is also greater as the water flowing in the river channel in the lower course does not have to overcome as much friction as that in the upper course, which has angular rocks and a shallow channel.

Most of the weathering of bare rock happens in mountain areas, where it is exposed. This material can then fall down the steep valley sides into the upper course of the river. It is still very angular, as the results show. As it moves downstream it hits the sides of the river bed, and also other rocks that make up the load. This knocks the sharp edges off the material, smoothing its sides and making it rounded. The load of the river, therefore is noticeably more angular in the upper course, but becomes rounded in the lower course – even on a relatively short river such as the Glendun.

Figure 19
Measurements being taken on the Glendun River.

Processes carried out by a river – erosion, transportation and deposition

Erosion

When rivers have a large bed load made up of coarse material they scrape or rub against the channel bed, eventually lowering the level of the bed, creating steep valley sides. This is vertical (downwards) erosion.

In sections of the river channel where the river is flowing especially fast, the water itself has enough energy to wash away the bank of the river, leading to undercutting and collapse. As this is a sideways motion, it is called lateral erosion.

Transportation

All rivers contain minerals and solid material, known as the load of the river. Weathered material falling into the river from the valley sides forms 90% of the load. The remaining 10% is the result of erosion caused by the river of its own banks and bed.

Rivers move their load in four ways:

1 Traction – the rolling of large rocks along the river bed. This requires a lot of energy, and the largest bed load will only be moved like this in times of severe flood.

2 Saltation – the bouncing of medium-sized load along the river bed.

3 Suspension – the smallest load, like fine sand and clay, is held up continually within the river water. This makes the water appear opaque. Some rivers carry huge quantities of suspended material, for example the Yellow River in China has enough sediment suspended in its flow at any one time to bury the city of London a metre deep.

4 Solution – soluble minerals dissolve in the water and are carried in solution. This may also colour the water, for example water in the rivers of the Mournes often appears yellow/brown as it is stained from iron coming off the surrounding peat bog.

Deposition

When the velocity of the river is reduced, its energy falls, and it can no longer erode or transport material, instead, the load is dropped, starting with the largest, and therefore heaviest, particles. This process is called deposition.

Conditions when deposition is likely:

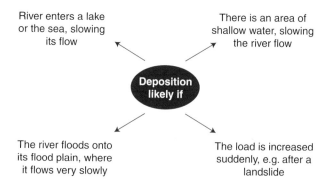

River enters a lake or the sea, slowing its flow

There is an area of shallow water, slowing the river flow

Deposition likely if

The river floods onto its flood plain, where it flows very slowly

The load is increased suddenly, e.g. after a landslide

It is the combination of erosion, transportation and deposition that create the general landforms seen along a river channel.

Formation of fluvial features

If you are investigating rivers in your coursework try using photographs to bring life into the project, but it is important to use them appropriately.

When using photographs:

- Only photograph what is relevant to the project – give it a full title.

- Focus in on the important feature or building you want. Annotate the photograph in the final project.

- Include an object such as a metre rule to give an idea of scale.

Remember to keep a record of what each photograph shows. In this case it was Station 1 on the Glendun River.

Field sketches are another method of illustrating project work. Here you draw a simplified picture of the geographical feature or area you are studying.

Figure 20
Methods a river uses to move its load.

1   Traction

2   Saltation

3   Suspension

4   Solution

activities

1 (a) Using a spreadsheet package, draw a graph to illustrate one of the Glendun River's characteristics. Use the information on page 45 to help you [5]

(b) Describe and explain the trend shown by the graph. [7]

Top Tip

Don't forget to use figures when describing a graph.

2 (a) State what is meant by the following terms: discharge and load. [4]

(b) Describe five safety measures you would need to take if you were carrying out a river study. [10]

3 (a) Name the fluvial feature found in the map on page 44 in grid square 15 23. [2]

(b) State three facilities tourists might use in the map areas shown, and give 6-figure grid references for them. [9]

(c) State the distance the river covers in its middle course. Give your answer in kilometres. [2]

HIGHER TIER

4 (a) Name and describe four methods that rivers use to move their load. [12]

(b) Explain why rivers move more load in the winter time than the summer time. [5]

(c) Using Fig. 18 on page 44, draw a cross-section of the river in its upper course in grid square 16 25 and in its lower course in grid square 23 32. Use these to help you describe and explain whether the river is eroding vertically or laterally at these points. [15]

Figure 21
Landscape features in a drainage basin

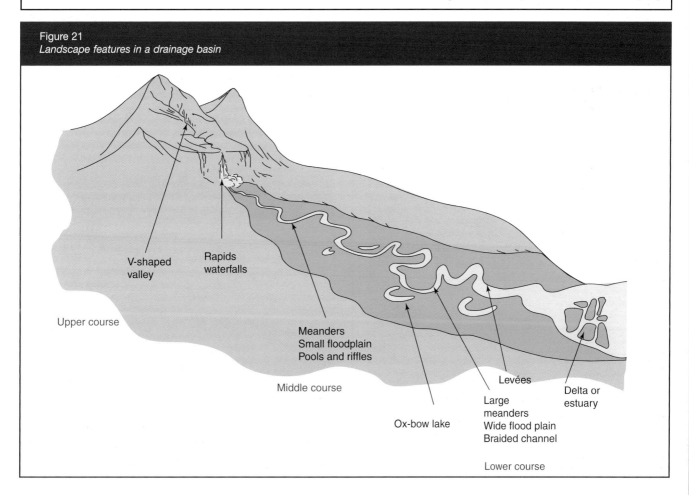

V-shaped valley

Rapids waterfalls

Upper course

Meanders
Small floodplain
Pools and riffles

Middle course

Ox-bow lake

Large meanders
Wide flood plain
Braided channel

Levées

Delta or estuary

Lower course

Figure 22
Station 1 on the Glendun River.

Each field sketch should then be given an appropriate title and be fully labelled (see Fig. 23).

Figure 23
Drawing a field sketch of Station 1 of the Glendun River.

1st stage:

Draw a frame the size you want your field sketch to be, then add the main lines, separating different land uses, water from land, or important buildings

2nd stage:

Add colour to clarify features

3rd stage:

Add annotation

Steep sides of v-shaped valley

Narrow river channel

Grasses on valley sides

Large bed load

Rapids with 'white water'

Meanders

Meanders are bends that develop in a river channel as the gradient (slope) of the river evens out. They are continuously changing features that are the result of differences in the velocity of the river across its channel. Where water flows fastest in the channel it spirals downwards, causing vertical erosion, deepening the river channel and creating a river cliff on the bank. Opposite this, water flows very slowly and does not have enough energy to erode. It cannot even hold up the load it is carrying, so it drops (deposits) the heaviest material first, then the next largest and so on, until only the smallest clay particles may be left in suspension. This leads to a lop-sided cross-section through a meander – see Fig. 24 below.

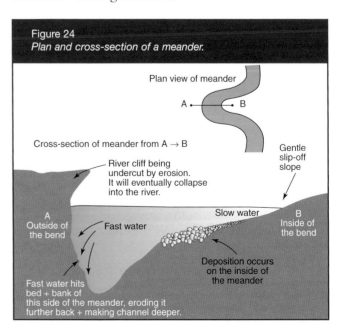

Figure 24
Plan and cross-section of a meander.

Plan view of meander

A • → • B

Cross-section of meander from A → B

Gentle slip-off slope

River cliff being undercut by erosion. It will eventually collapse into the river.

Slow water

A
Outside of the bend

Fast water

B
Inside of the bend

Deposition occurs on the inside of the meander

Fast water hits bed + bank of this side of the meander, eroding it further back + making channel deeper.

In the middle and lower courses of a river, meanders are constantly being formed and reformed. The bends can get bigger and sometimes they can even be cut off altogether, and an ox-bow lake is formed.

A very good example of a river that has clear ox-bow lakes and meanders is the Mississippi River in the USA. This can be seen at the website www.topozone.com.

Floodplain

As the river meanders back and forth, it flattens the land around it creating a floodplain either side of the river in its valley. The floodplain is covered in sediments that have been deposited by the river in times of flood. This material is called alluvium, and is very fertile, which is why river valleys make good places to grow crops. Floodplains can be easily recognised on maps. Look out for the features shown in Fig. 25 on Ordnance Survey maps.

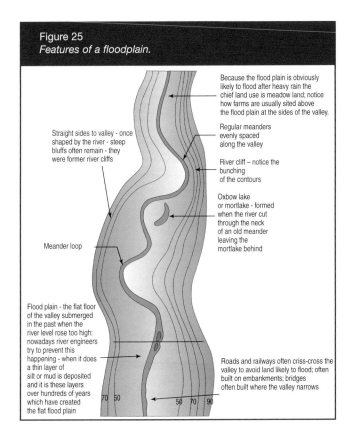

Figure 25
Features of a floodplain.

Straight sides to valley - once shaped by the river - steep bluffs often remain - they were former river cliffs

Meander loop

Flood plain - the flat floor of the valley submerged in the past when the river level rose too high: nowadays river engineers try to prevent this happening - when it does a thin layer of silt or mud is deposited and it is these layers over hundreds of years which have created the flat flood plain

Because the flood plain is obviously likely to flood after heavy rain the chief land use is meadow land; notice how farms are usually sited above the flood plain at the sides of the valley.

Regular meanders evenly spaced along the valley

River cliff – notice the bunching of the contours

Oxbow lake or mortlake - formed when the river cut through the neck of an old meander leaving the mortlake behind

Roads and railways often criss-cross the valley to avoid land likely to flood; often built on embankments; bridges often built where the valley narrows

Levées

When the river does overflow its banks it quickly loses energy and so must deposit much of its load onto the floodplain. As the largest load is deposited first it quickly builds up to form natural embankments called levées.

During low flow, the river may also deposit material on its bed, if the velocity becomes very slow as the amount of water in the river falls. The river might

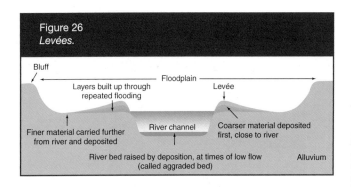

Figure 26
Levées.

Bluff

Floodplain

Layers built up through repeated flooding

Levée

Finer material carried further from river and deposited

River channel

Coarser material deposited first, close to river

River bed raised by deposition, at times of low flow (called aggraded bed)

Alluvium

even dry up altogether. If load is deposited onto the river bed, and not washed away later in the season, the river bed can be raised and in some cases the river may end up flowing above the level of the floodplain.

Levées can be artificially strengthened and raised to protect the floodplain against flood damage.

Deltas

A delta is a feature that is found at the mouth of some rivers – where the river meets the sea. They are features made by deposition, so are more obvious on long rivers that have gathered a large load. As the river reaches its lower stretches it deposits almost all of its load. This is because it slows down dramatically as it meets the sea – a larger body of relatively still water. Also, particles of clay in the river water join together to make larger, heavier particles due to the mixing of fresh and salt water. So long as the deposition occurs at a faster rate than the tidal changes remove material out to sea, it can build up and break through the surface of the river, forming islands, around which the river must now flow.

The coarsest material is deposited first and so is found closest to the coastline.

There are two main types of deltas depending on their shape:

- Birds foot delta – where fingers of deposited material extend out into the sea. The Mississippi has this type of delta.

- Arcuate delta – a triangular-shaped delta, that doesn't extend into the sea. It often has a smooth coastline. The Nile has an arcuate delta.

Where the tidal currents remove some of the deposited load out to sea, it can only build up along the sides of the river at its mouth, creating an estuary. The alluvium can break through the surface and form mudflats, which are an important habitat for many wading birds.

Figure 27
Features of a delta.

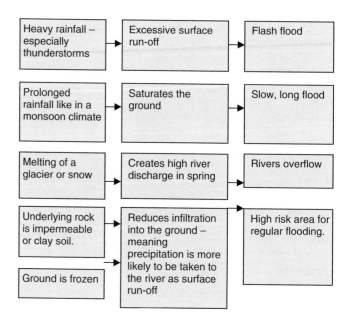

Sea

Distributaries

Meandering river

Flood plain

activities

1   (a) Using the map extract of the Glendun River, draw a sketch map showing the river, and annotate all of all the fluvial features.   [10]

    (b) Describe and explain the formation of one of the features you have labelled.   [7]

2   (a) Describe how levées form. Use a series of diagrams to help you.   [8]

    (b) State fully one reason why levées are important features to people living on floodplains.   [3]

3   Name and describe the two main types of river deltas.   [6]

4   **HIGHER TIER**

Research deltas using the Internet and create a PowerPoint report on one delta. Include its name, a photograph and some information on its formation and its future. Large rivers with deltas include the Nile, Mississippi, Yellow River, Yangtze, Rhine, Ganges and Amazon.   [10]

web link & extra resources

whttp://www.athenapub.com/rivmiss2.htm – this site has a satellite image of the Mississippi Delta.

http://www.lacoast.gov/cwppra/projects/mississippi/index.asp – a good site that covers conservation projects and information on the Mississippi Delta.

http://en.wikipedia.org/wiki/Main_Page – a free encyclopedia for you to investigate any delta or river.

FLOOD HAZARD

Rivers can serve us in many ways:

- a source of water – **irrigation** and water supply
- a source of water power – hydro-electricity or water mills;
- a waste outlet – used as a drain taking effluent out to sea;
- a routeway – roads and rail links follow valleys, river traffic and shipping;
- a source of food – fish makes up the main source of protein in many LEDCs.

Water is a vital resource, but too much water can be a hazard.

Floods are temporary excesses of water that cover areas that are usually dry. Flooding forms part of the natural annual pattern of discharge of most rivers. Flooding can happen to low-lying areas, whether or not a river is nearby. They are also occur frequently after storm surges in coastal areas. This chapter concentrates on floods caused by rivers bursting their banks. In other words, rivers that have exceeded their banks' full level, and so spill out of their channel on to the surrounding landscape.

Below is a summary for Higher Tier students of the causes, impacts and flood prevention methods. These are developed for all in the following case study based on the Mississippi River.

Causes of flooding

Physical causes

Heavy rainfall – especially thunderstorms	→ Excessive surface run-off	→ Flash flood
Prolonged rainfall like in a monsoon climate	→ Saturates the ground	→ Slow, long flood
Melting of a glacier or snow	→ Creates high river discharge in spring	→ Rivers overflow
Underlying rock is impermeable or clay soil.	Reduces infiltration into the ground – meaning precipitation is more likely to be taken to the river as surface run-off	→ High risk area for regular flooding.
Ground is frozen		

Human causes

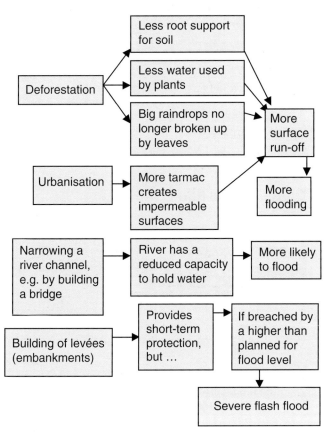

Impacts of flooding

Positive impacts:

- Replenishes drinking water supplies especially wells.

- Provides water to irrigate crops, e.g. rice that needs to grow in wet conditions.

- Provides sediment (other terms are silt or alluvium) that naturally fertilises the soils of the floodplain.

- Provides a habitat for many species of fish to breed and shellfish to live, these can be gathered as a food source.

Countries such as Bangladesh and Egypt rely on flood.

Negative impacts:

- Spreads water-borne diseases.

- People and animals can drown.

- Infrastructure (roads and railways) may be disrupted.

- Buildings can be damaged or destroyed.

- Crops grown on fertile floodplains can be washed away in a flash flood.

- Land may become waterlogged, so cannot be farmed.

- People can be made homeless.

River management schemes aim to control rivers and reduce the risk of unwanted flooding. Planners can respond to the flood hazard by changing the river through engineering. They might implement hard, or soft, flood control measures.

Soft engineering flood control

These are generally sympathetic to the natural landscape, so tend not to damage the river for future generations, making them more sustainable than hard flood control measures.

They may involve:

- Planting trees (afforestation) in the upper course of the river;

- Planting water-loving plants in the floodplain. These soak up water, lowering the water table;

- Enlarging existing levées, or creating embankments set back from the river. The latter will still allow some natural flooding onto the floodplain.

Hard engineering flood control

They often involve making large man-made structures to control the river, breaking its natural cycle of flood and subsidence. These measures are not sustainable in the long term.

They may involve:

- Building a dam or reservoir in the upper course. The resulting reservoir can be used for leisure and hydroelectricity, but can flood good farm land and displace local people and destroy habitats;

- Changes to the river channel. By deepening and widening the river channel, they increase its cross-sectional area, allowing it to contain more water, meaning the discharge has to be greater to create a food;

- Building high embankments along the sides of the river to contain any floodwater.

CASE STUDY

A RIVER MANAGEMENT SCHEME AT THE NATIONAL LEVEL – THE MISSISSIPPI

Figure 28
Main features of the Mississippi River.

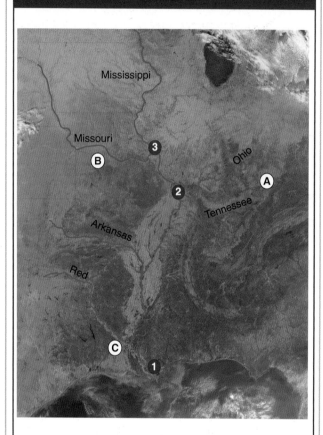

(A) Ohio–Tennessee rise in the Appalachians which receive heavy cyclonic rainfall between January and May. Flood risk increase following snow melt.

(B) Right-bank tributaries drain relatively dry. Mid-west rainfall mostly falls in summer when evaporation is at its highest.

(C) Lower Mississippi usual and most frequent floods.

(1) New Orleans

(2) Cairo

(3) St Louis

The Mississippi together with its main tributaries, the Missouri and the Ohio, drains a third of the USA and a small part of Canada. Altogether the total area drained by the watershed is 1.6 million square miles (remember that Northern Ireland is only 5456 square miles). Left to follow its natural cycle, flooding would be an almost annual event with late spring having the peak discharge. The Mississippi is the fourth longest river in the world.

It is a very important river to the USA. A study in 2000 estimated that 18 million people rely on this river for their water supply – this includes 50 cities.

Causes of the flood

MISSISSIPPI

FLOOD 2001

Why does the Mississippi flood?

Flooding is an act of nature. It can become a hazard because people often choose to live on floodplains. Without human intervention, the Mississippi would flood almost every year, although some recent floods have been extremely severe.

In 2001 the 22.5 foot peak discharge was the third highest on record.

Experts say record-setting wet weather was the main culprit. There is usually heavy rainfall from January to May in the Appalachian Mountains, especially if it happens at the same time as snowmelt in the mountains. This excess water raises discharge to a level that can cause flooding. There was wet weather before the 2001 flood.

Many floods occur in the lower course below Cairo and Illinois because the land flattens out.

The main floodwater comes from the Ohio and Tennessee due to the heavier rainfall and spring snowmelt in the Appalachians. The mid-west source region of the Mississippi is semi-arid so is not normally a source of floodwater.

Although weather remained the main cause of the floods in 2001, recurring bouts of high level floods can be seen in part as a response to human interference in the drainage basin system.

Straightening the channel contributes to flooding. Rivers try to create meanders to balance their flow. Without them water flows very quickly down the channel, creating wave crests that can be higher than the main flow, breaking over artificial levées.

More towns and cities have been built along the banks of the Mississippi and on its floodplain, this means more impermeable surfaces and therefore more rapid run-off.

continues

continues

Many areas of wetland have been drained around the river. These acted as natural stores for floodwater. Since work began on the Mississippi, 40 million hectares (ha) of wetland have been lost.

Ploughing up prairie grass and planting crops in rows allows more surface run-off, which ends up as discharge in the river.

Impact of flooding on people and environment

MISSISSIPPI FLOOD 2001

April 15, 2001

It's official; River hits flood stage – News Reports from 2001

The river swelled above the 15-foot flood stage at Lock and Dam 15 in the Quad-Cities Sunday, hitting 15.16 feet by 9:30 pm. It's not expected to peak until April 24 or 25, when the National Weather Service and the US Army Corps of Engineers predict it could hit the 22-foot mark.

That's just shy of the record-setting 22.63 foot crest of July 9, 1993, and the runner-up 22.48 crest of April 1965.

Swelled by rapid snowmelt and heavy rains across the Upper Mississippi basin, the river was rapidly rising from Guttenberg south. In the Quad-Cities, the river rose nearly a foot between Saturday and Sunday.

Those living near the river's banks continued working Sunday to sandbag around their houses and move belongings to higher ground. Businesses along the waterfront in Davenport, such as the Rhythm City Casino, were also busy building sandbag dikes.

Interviews with locals show how the flood has affected people.

Positive impact

Farmer

'The main positive impact is that frequent flooding over the centuries has allowed today's Mississippi to flow for much of its middle and lower course over a wide, flat alluvial floodplain. The soil here is very fertile because of the silt (alluvium) that the Mississippi deposits when it floods. The soil we have here is called Loess, and it can support a wide variety of crops, including soybeans, corn (maize), cotton and wheat. Unfortunately, if there is a flash flood in the spring, it can wash seedlings out, and my crop can be destroyed.'

Town Planner, Davenport

'The flat land alongside the river is easy to build on. It costs less to build here than to have to meet the expense of flattening out hills to create a foundation for housing or factories. It has been flooding over the years that has flattened out the land in the valley.'

Owner of Natchez Steam Boat

I couldn't run my business without the river. The Mississippi has a gentle gradient for river transport, like the river boat and has a long deep channel that allows larger boats to sail far inland, which is good for importing and exporting goods. However, I run a pleasure boat service where people can sail up the Mississippi from New Orleans. I charge a minimum of $65 for a day trip.'

Bird Watcher

'Along the Mississippi River shores in St. Paul and Minneapolis, you almost feel the presence of fur traders, explorers, buffalo hunters, soldiers, settlers, native Americans and riverboat captains that travelled this area many years ago. The floodplain, with its natural marshes, is home today to much wildlife including, beaver, otter, fox, hawks, egrets, herons and the majestic bald eagle.'

Negative impact

Man who works for Alter Barge Company

'The Army Corps of Engineers have closed a large section of the Mississippi to barge traffic. This means over 2000 of the barges we move, packed with fertilisers and grain, cannot make it to their delivery points down river from Iowa. We can't do anything about it, but we are losing $50,000 a day.'

City Administrator in Davenport City

'We've had everything. The river has flooded, we've had our tornado warning and a flash flood, all we need now is an earthquake. We don't have the man-power to cope. Eventually the cost will be passed on to our tax-payers, so far our damage is just over $1 million, but it could still rise.'

Nurse

'I am teaching people how to avoid the health risks associated with these flood waters. 'Soap and water' or steering clear of floodwaters is the best defence of water-borne illnesses such as Hepatitis A and typhoid. Don't drink the water. If people are worried about their drinking water, I am advising them to boil it for at least 2 minutes.'

Teenage girl

'We live on Campbells Island, but on Tuesday the flood meant we had to be evacuated. It was too dangerous to cross the Mississippi in our family boat, because the current was too strong, we would have been washed away! It was a bit of an adventure, to be taken off the island by helicopter, but now we are spending the week in temporary shelter.'

Management response – hard and soft engineering projects

Preventing flood-related damage can be accomplished by building structures, such as reservoirs, levées, channels and floodwalls, that change the characteristics of floods. In addition, non-structural measures, such as floodplain

continues

continues

evacuation and flood proofing help reduce the flood risk to humans.

The table below shows some of the main responses to the flooding of the Mississippi.

Hard engineering	Soft engineering
Main policy has been to hold back with levées. Levées have been raised in places to over 15 m and strengthened. There are about 3000 km of levées along the Mississippi and its tributaries	Afforestation to reduce and delay runoff, e.g. in the Tennessee Valley
By cutting through the meanders the Mississippi has been straightened for 1750 km, where it flows through artificial channels, looking more like a canal.	Diversion spillways to divert flood water. This is carried out in the delta – water is therefore moved quickly off the land and out to sea.
The flow of the major tributaries (Missouri, Ohio, and Tennessee) have been controlled by a series of dams – over 100 in total!	Creation of safe flooding zones, by buying housing in the current flood danger areas and creating green areas instead. In Rock Island, Illinois, $7 million worth of housing has been purchased by the County

Most of these engineering projects are funded by the Mississippi River and Tributaries Project (MR+T).

Need for a co-ordinated approach to ensure sustainable development

The Mississippi River is one of the largest in the world, and 31 states of the USA rely on it for some of their water supply, all of whom are therefore interested in changes made to the river.

The 1993 floods made river control a priority because of the compensation costs. During that flood nearly half the counties in the nine states bordering the upper reaches of the Mississippi and Missouri rivers were declared federal disaster areas and therefore were entitled to Federal Aid.

Since all of the people in the USA would have to fund compensation given as Federal Aid in the event of massive flood damage, everyone in the USA has a vested interest in making sure as little flood damage is done as possible. In 2002 the government of the USA invested $10.8 billion in MR+T, but estimated that the amount saved from flood damage claims that would have been filed if there had been no flood prevention schemes to be $258 billion. Therefore, these schemes are cost-effective for the US government.

The Army Corps of Engineers are responsible for physically building and maintaining flood control in the basin – all states through which the river and its tributaries flow must work with this authority to co-ordinate flood control.

So, is the Mississippi being controlled in a way that allows the present generation of people and wildlife to meet their needs without endangering future generations to meet their needs? In other words, is all this river control sustainable?

The Mystery of the Vanishing Delta

Up until 1950 the Mississippi Delta was growing at the rate of a few square kilometres a year. In the last 50 years this trend has been reversed. The wetlands are disappearing and the shore of the delta is retreating by as much as 25 metres per year.

Deltas grow as long as there is enough sediment reaching them in the river. The levées that have been built along the Mississippi have been altering the rivers natural balance for a long time now – the first levées in New Orleans were built in the 18th century. Together with the spillways and jetties at the mouth of the river, water is kept moving fast and routed straight out to sea, bypassing the natural mouth. All this has stopped the build up of sediment needed to maintain the delta area.

The problem is compounded by soil conservation measures in the source region of the Mid-west and rising sea levels. It looks like the Mississippi Delta is vanishing right before our very eyes!

For wildlife, any interference to the natural river system is damaging. The draining of so many marsh areas and the reduction, indeed in some places the complete cessation, of flooding means habitat loss that may lead to their endangerment or even extinction. Such animals include herons.

For the people who live along the river it is also unsustainable because the silting of the river channel behind dams means that it takes less discharge to cause a flood and as levées rise to

continues

continues

hold the river at a level above the floodplain, any breach of the levée can cause a devastating flash flood. Parts of New Orleans are now more than 4.3 metres below river level.

Figure 29
Most of these engineering projects are funded by the Mississippi River and Tributaries Project (MR+T).

web link & extra resources

www.big-river.com/ – a good general site on the Mississippi.

www.mshistory.k12.ms.us/features/feature25/msriver.html – study the history of the Mississippi.

www.qconline.com/rivers/ – the Quad-Cities website. You can see live webcams of the river and even get current stream flow data from gauging stations along the Mississippi.

www.mvr.usace.army.mil/PublicAffairsOffice/DavenportFloodControl.htm – follow flood control developments for Davenport.

activities

1 (a) Describe three conditions under which a river is likely to flood. [6]

   (b) Explain the difference between hard and soft engineering measures to protect against flooding. [6]

2 (a) Name a river you have studied that has flooded, and give the year it flooded. [2]

   (b) State fully one positive and one negative impact the flood had. [6]

   (c) Describe one management response to the flood hazard presented by your named river. [3]

3 (a) State fully one physical and one human cause of the Mississippi flood of 2001. [6]

   (b) Explain why a large river, such as the Mississippi, must have a co-ordinated approach to sustainable river development. [6]

4 Draw a map of the Mississippi and label important information about the impact of flooding onto relevant parts of this map. [9]

HIGHER TIER

4 (a) With reference to a river management scheme you have studied, describe the hard and soft engineering measures used and evaluate their effectiveness. [12]

   (b) Look at a map of a section of the lower course of the Mississippi. Draw an annotated sketch of the fluvial feature it shows and write a paragraph explaining how it was formed. [6]

   (c) Not all of the Mississippi is heavily managed. One such area is the town of Davenport. In 2001 they faced a decision, to build a flood protection wall or live with flood. Using the internet site posted on Davenport, imagine you are a local Davenport resident and write a speech to give at the next council meeting where they will be discussing whether Davenport should build a flood protection wall. Ensure you include at least two arguments to back up your chosen position. [10]

UNIT THREE
Limestone Landscapes and Their Management

ROLE OF ROCK TYPE, STRUCTURE AND WEATHERING

Rocks are classified by the manner in which they were made, how they have been changed and what they are made from.

Igneous rocks

These are formed when molten lava or magma cools and hardens. If the lava has been exposed on the surface, it may cool quite quickly, producing few if any crystals within its structure. The Giant's Causeway in County Antrim is made of one such stone – Basalt. If the igneous rock is made from magma that cooled slowly underground, then large mineral crystals form, and speckled igneous rocks form. The granite that is found in the Mourne Mountains in County Down was created in this manner.

Sedimentary rocks

Weathering and erosion of rocks produces sediments, small fragments or particles, which accumulate on land, coasts and marine environments. Over time, layers of these fragments form on top of one another, causing the air and moisture to be squeezed out, and a solid rock to be created. These are sedimentary rocks. The line between layers of sediment is a line of weakness called a bedding plane.

As well as being made from fragments of rocks, sedimentary rocks can be made from the chemicals left after the evaporation of water, or from layers of plant and animal remains. Occasionally, plant and animal remains do not get crushed by the process and remain trapped in the rock as a fossil.

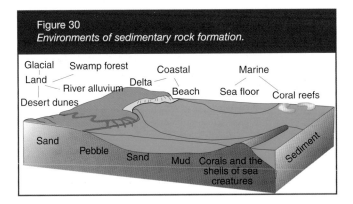

Figure 30
Environments of sedimentary rock formation.

Glacial Swamp forest   Coastal   Marine
Land ____ River alluvium Delta ___ Beach Sea floor Coral reefs
Desert dunes

Sand   Sediment
Pebble   Sand   Mud   Corals and the
                        shells of sea
                        creatures

Metamorphic rocks

The final main group of rocks are the metamorphic rocks. These are rocks that have been altered or changed by extreme heat or pressure. They were once either igneous or sedimentary rocks. Sometimes the pressure and heat have changed the rock so much that it can be very hard to tell what it originally was.

We can do simple experiments to investigate the origin of metamorphic rocks. Marble is a metamorphic rock that fizzes when acid is poured onto it, just like limestone fizzes under the same experimental conditions. Further chemical analysis of the two types of rock show that they are both mostly made from calcium carbonate. The fragments of shell that created the limestone have been crystallised by heat to make marble.

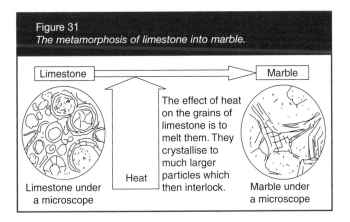

Figure 31
The metamorphosis of limestone into marble.

Limestone → Marble

The effect of heat on the grains of limestone is to melt them. They crystallise to much larger particles which then interlock.

Limestone under   Heat   Marble under
a microscope   a microscope

Weathering

Weathering is the breakdown of rocks by the weather, without the movement of the rock itself (in situ). It can happen in three ways – mechanical, chemical and biological.

Mechanical weathering – peeling

Changes in temperature can cause rocks to expand when heated and contract when cooled. If an area has a wide daily temperature range, such as a desert, then the continual and regular expansion and contraction can weaken the rock structure, and the outer layer of the rock can fall off, a process called onion peeling. Some rocks, like basalt, have an internal structure that is prone to this type of weathering and interesting features may be produced by onion peeling, such as the giant's eyes seen at the Giant's Causeway.

Mechanical weathering– frost shattering

This is also caused by temperature changes, but rather than directly affecting the rock, water is involved. Rocks have cracks on the surface that may collect water after rainfall. When the temperature falls below freezing the water in the crack freezes and expands. If water is regularly frozen in this crack, the crack will start to widen, allowing even more water in. Over time, this crack can become so wide that it breaks the rock into pieces. In granite landscapes it leads to the formation of tors.

Chemical weathering

Rainwater and river water can have dissolved chemicals in it, which over time will in turn react with rocks, changing them chemically, or even dissolving them away. As chemical reactions increase with increased temperature, chemical weathering is greatest in hot, humid environments.

Biological weathering

Plants and animals can also break down rocks. The roots of plants can penetrate cracks, and widen them as the roots expand and grow. Also, plants can exude chemicals that may aid weathering of the rock. Burrowing animals, like rabbits and worms, also break up rocks as they excavate tunnels. Remember we do not have moles in Northern Ireland, so they do not contribute to the biological weathering of rocks here.

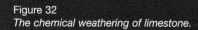

Figure 32
The chemical weathering of limestone.

H_2O — Rain

and

+ — Soil water

combining with

CO_2 — Carbon dioxide from the air and from organic material in the soil

Creating

H_2CO_3 — Dilute acid (Carbonic acid)

+ — when this meets

$CaCO_3$ — Limestone (calcium carbonate)

Its dissolves the rock

$Ca(HCO_3)_2$ — The water then contains dissolved calcium carbonate (calcium bicarbonate)

When this limestone-loaded water meets cave air the chemical reactions are reversed

CO_2
+
$CaCO_3$
+
H_2O

The water loses some carbon dioxide to the air and the water becomes less acidic

This forces some calcium carbonate to be deposited as calcite

LIMESTONE STRUCTURE AND WEATHERING

Limestone mainly consists of calcium carbonate, which comes from the remains of seashells. It is a sedimentary rock.

Type of limestone	Colour, texture, permeability
Carboniferous limestone	Hard, grey, pervious
Jurassic limestone	Soft, yellowish, porous
Chalk	Soft, white, porous

Each type of limestone develops its own special type of scenery. Carboniferous limestone scenery is known as karst.

There are three factors that influence the development of karst landforms:

- The permeability of limestone – it is pervious or porous, so water can easily enter the rock to weather it.

- The vulnerability of limestone to chemical weathering from rainwater – calcium carbonate reacts with rainwater, which is mildly acidic, to form calcium hydrogen carbonate, which is soluble (see Fig. 32).

- The structure of carboniferous limestone. The bedding planes and joints are areas of weakness that are dissolved and widened by chemical weathering.

Figure 33
The structure of limestone.

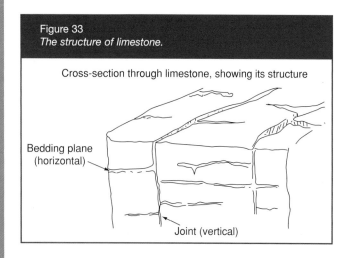

Cross-section through limestone, showing its structure

Bedding plane (horizontal)

Joint (vertical)

activities

1 Describe briefly how igneous and metamorphic rock types are formed. [6]

2 State the meaning of weathering. [2]

3 State fully how limestone is weathered by rainfall. [4]

HIGHER TIER

4 Explain how the structure of limestone is an important factor in how this rock is weathered. [6]

FEATURES OF LIMESTONE ENVIRONMENT

Ireland has a vast variety of geology. One region that has an outstanding limestone area is in Co. Clare, in an area called the Burren. The Burren covers 360 km² and forms a plateau gently tilted to the south, at 200–300 m above sea level in the north and 100 m in the south. The Burren (from the Irish *bhoireann* – a stony place) lies in the north-western corner of Co. Clare. It is the largest area of carboniferous limestone in western Europe.

Surface features

Swallow holes

Carboniferous limestone areas are characterised by a lack of surface drainage. When a river flows from an impermeable rock onto carboniferous limestone, the acids from the river water begin to dissolve and widen surface joints in the limestone to form a swallow hole. In time the river will disappear down one of these swallow holes and flow underground.

When the river meets an impermeable rock it is forced to flow over this rock until it reaches the surface. The place where it reappears is called a resurgence spring.

For example, in the Burren only at the junction of the limestone and the impermeable shale rocks, around Slieve Elva for example, are there valleys containing streams. Where these streams cross from

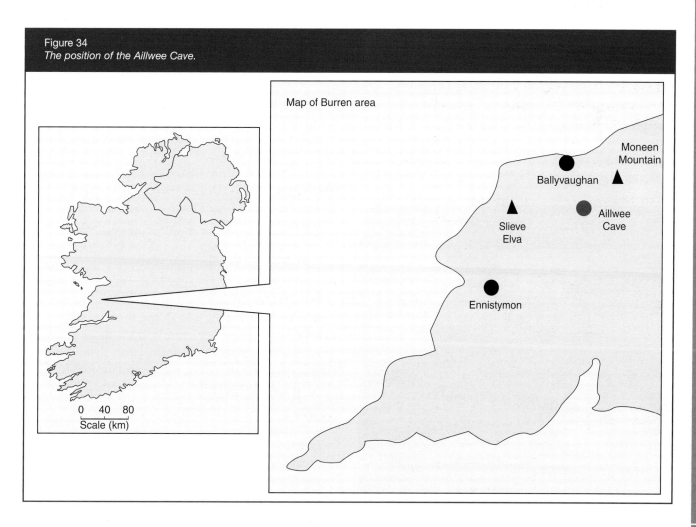

Figure 34
The position of the Aillwee Cave.

Map of Burren area

Moneen Mountain

Ballyvaughan

Slieve Elva

Aillwee Cave

Ennistymon

0   40   80
Scale (km)

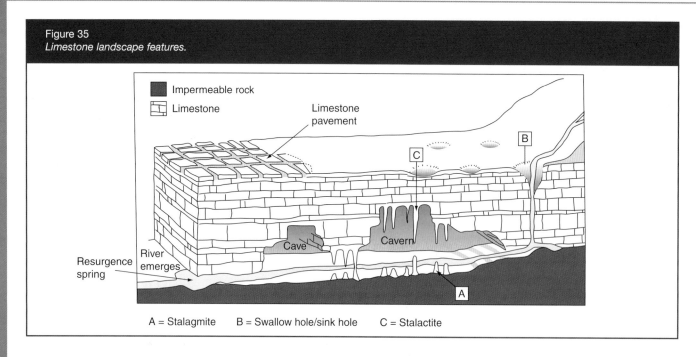

Figure 35
Limestone landscape features.

■ Impermeable rock
▨ Limestone

Limestone pavement

Cave

Cavern

Resurgence spring
River emerges

A = Stalagmite   B = Swallow hole/sink hole   C = Stalactite

the shale to the limestone they disappear underground at swallow holes, the waters flowing through cave systems before emerging from springs such as those at Killeany and St. Brendan's Well near Lisdoonvarna.

Limestone pavement

Limestone scenery usually only has a thin soil cover because calcium carbonate dissolves and does not provide much material to form the basis for soil.

In the Burren the limestone was exposed due to overgrazing by sheep in the Bronze Age. This led to extensive soil erosion, and eventually the exposure of the limestone bedrock. When there is no soil, the top bedding plane is exposed. The joints are widened by

Figure 36
Limestone pavement.

Clint

Gryke

solution to form grooves called grykes and the flat topped blocks between them are called clints.

The grykes provide a sheltered environment for wildlife. Many rare species of plants thrive in the Burren, living in these sheltered pockets. Plant species found in the Burren include fragrant orchid, spring gentian and the heath spotted orchid.

Underground features

Caves and caverns

Underground rivers widen joints and bedding planes as they flow along them. In some places solution is so active that underground caves form. Where a series of caves form they are linked by narrow passages – ideal for potholers. Rivers abandon these caves as they try to find a lower level.

Within the Burren, the most famous cave system is the Aillwee Cave. It was discovered by a local farmer when his sheep dog disappeared into what is now the entrance of the cave system. The remains of brown bears and their hibernation pits have been found within the caves. This was an important find, as bears have been extinct in Ireland for thousands of years. Now tourists can visit the caves that extend over a third of a mile into the mountainside. In the caves the temperature is a constant 5.5°C all year round.

About 1 million years ago the ice age began and from then until 15,000 years ago Ireland's climate alternated between arctic cold and warmer periods, freezing and melting over the centuries. This melting water roared and crashed its way through the Aillwee caves greatly enlarging the passage and bringing with it large quantities of sand and silts which are still present in the inner cave.

Aillwee is one of the most ancient caves in the Burren and perhaps in Ireland.

Stalactites and stalagmites

Figure 37
Stalactites and stalagmites in the Aillwee.

Cave structures are formed by the deposition of calcite dissolved in groundwater. Stalactites grow downwards from the roofs or walls and can be icicle-shaped, straw-shaped, curtain-shaped or even form terraces. Stalagmites grow upwards from the cave floor and can be conical, fir-cone-shaped or resemble a stack of saucers. Stalactites and stalagmites may meet to form a unbroken column from floor to ceiling.

Stalactites are formed when groundwater, hanging as a drip, loses a proportion of its carbon dioxide into the air of the cave. This reduces the amount of calcite that can be held in solution, and a small trace of calcite is deposited. Successive drips build up the stalactite over many years. In stalagmite formation the calcite comes out of the solution because of the disturbance caused by the shock of a drop of water hitting the floor removes some calcite from the drop. The different shapes result from the splashing of the falling water.

Top Tip

Remember, stalactites (with a 'c' in the middle) grow from the ceiling of the cave, and stalagmites (with a 'g' in the middle) grow up from the ground. It might also help to think that since they are hanging from the ceiling, stalactites must hold on tight.

It takes a long time for the deposits of calcite to grow into large structures. For example, over a hundred years, a stalactite may grow one cubic centimetre down towards the floor.

The larger stalagmites in Midsummer Cavern in the Ailwee caves took 5000 years to form. Poll-an-Ionain, near Lisdoonvarna, is a difficult cave to explore and has inside it a huge stalactite nearly 25 feet long – the longest in western Europe! Some samples of calcite taken from deep inside the Ailwee Cave, started to form 350,000 years ago.

activities

1 (a) Limestone is a sedimentary rock. Define the term sedimentary rock and describe two ways such rocks are created. [6]

   (b) Limestone pavements are a feature found in karst areas. Describe and explain their formation. [6]

   (c) Other than a limestone pavement, name two other surface features of karst. [2]

2 Draw an annotated sketch of a cave system. It must include a swallow hole, cavern, stalactite, stalagmite and resurgence. [7]

3 Create a flow diagram to help explain how stalactites are formed. [6]

HIGHER TIER

4 Imagine you are a drop of rainwater about to fall on a karst area. Describe your journey from sky to sea – you might do this in the form of a cartoon strip and add humour. [10]

5 Using internet sites, create a tourist information leaflet on the Burren, pointing out the important karst features and explaining simply how they were formed. [10]

HUMAN PRESSURE ON THE LIMESTONE ENVIRONMENT

Limestone is a non-renewable resource, so must be carefully used in a sustainable manner if future generations are to enjoy the scenery it creates. The Burren has international importance and must be protected since only 2900 ha of limestone pavement now remain in Britain, 97% of which has already been damaged.

People make many uses of the limestone environment. Many of these impact negatively on the area, or cause problems for other users of the same area. Conflicts of interest arise from these different uses and sometimes something has to be done to resolve the conflict, or reduce the negative impact.

What is the situation in the Burren?

The main activities that are causing excessive pressure on the limestone environment of the Burren are tourism and quarrying. The current biodiversity is also under threat due to undergrazing. Study the information below.

For years flagstones have been quarried in the Moher area and the cutting of the stones has provided local employment. The limestone is used as paving around modern buildings, as well as for fireplaces and other features.

The Irish Government has now legislated against the removal of limestone from certain areas of the Burren. It is also protected under European Law – limestone pavement is protected under the European Habitats Directive (92/43/EEC). They have proven the laws are enforceable, and must be followed. Indeed, a man from Tubber was sentenced to four months imprisonment in 1999 for removing limestone from the Burren. At Corofin District Court, Mr Toomey said it seemed as if the Department wasn't prepared to take stringent measures to safeguard the Bunratty site.

Figure 38
Quarrying in the Burren.

Cause of pressure on the limestone area	Impacts	Groups in conflict	Management response
Tourism – walking	Accelerated erosion of the limestone. Picking of rare wild flowers. Litter.	Tourists Environmentalists Farmers	Fenced off paths. Visitor centres now educate visitors about the importance of preserving the local flora.
Potholing	Accelerated erosion of the caves.	Pot holers Environmentalists	Restrict access to cave systems.
Quarrying	Removal of the limestone – non-renewable resource. Reduction of habitat for plants and animals of the area.	Quarry owners Environmentalists Local people Tourists Gardeners	Government can make quarrying from certain areas illegal. Gardeners should be educated about their choice of stone for walls and paths.
Farming	Overgrazing or undergrazing.	Farmers Environmentalists	Subsidies can be granted to farmers who do not overgraze to compensate them for the loss of profit experienced if fewer animals are kept on their land. Subsides can also be used to encourage framers to keep grazing animals, reducing undergrazing.

The Burren under threat

Many of the upland grasslands in the Burren are not being grazed sufficiently to uphold the high species diversity and withstand the advance of scrub.

The Burren upland region, Co. Clare is one of Europe's most precious and highly valued landscapes. This region contains almost three-quarters of Ireland's native plant species, like the bloody cranesbill, and boasts a remarkable collection of monuments including 500 ring forts.

Though often overshadowed by this priceless heritage, several hundred farm families also live in the Burren. For many generations these farmers have played a central role in creating and sustaining this rich environment.

Farming traditions

The Burren's upland grasslands are most commonly associated with the ancient tradition of winter grazing hardy native breeds of cattle, like the Shorthorn. This winter grazing promotes biodiversity through the removal of dominating grass and weed species, with minimal adverse impact on the other flora which can prosper unhindered over the summer. Vegetation studies found that winter grazing supports a plant species diversity 25% above that found in similar areas grazed mainly in summer. However, if grazing is relaxed on these grasslands, grasses, weeds and eventually scrub take over, ousting the native flora and resulting, on average, in an 18% loss in plant species diversity.

Farming today

Today many of the hardier cattle breeds have been replaced with more commercial breeds. Often they are used for breeding, and the additional nutrition needed by pregnant cows cannot be supplied by grass alone. More of these cows are being feed in cattlesheds with silage.

As a result, many upland grasslands are not being grazed sufficiently to uphold the high species diversity and withstand the inexorable advance of scrub – largely hazel – that now threatens to engulf stretches of this heritage-rich landscape.

How will we deal with the problem?

- The winter grazing season has been extended to run from 1 October to 30 April.
- Government subsidies to compensate for using less profitable farming techniques.

There has been much debate in the Burren area about the site for a new visitor centre …

Planning permission refused for Burren Centre

Monday, March 06 2000

Filed at: 11:50 PM

The Burren

Visitor centre not to go ahead

There has been a mixed reaction to the decision of An Bord Pleanála to refuse planning permission for a visitor centre at the Burren. The Minister for Heritage, said that she was disappointed at the decision. The Burren Action Group, which campaigned against the centre, said today's ruling had justified the stand they had taken for the past nine years. The conservation group, An Taisce, also welcomed An Bord Pleanála's ruling against the siting of the controversial visitor centre at Mullaghmore in the Burren National Park in County Clare.

The latest plan was a scaled-down version of the original. Upholding a decision by Clare County Council not to grant planning permission for the centre, An Bord Pleanála said that it would have led to unacceptable degradation of the environment and would detract from the scenery and rural character of the area.

Sadbh O Neill of An Taisce said that the decision means the Burren will be protected for future generations. This is possibly the last decision affecting the disputed visitor centre at Mullaghmore. The Minister has said that she still intends to press ahead with discussions on finding alternative visitor access to the Burren.

The group that had supported the project say that they are disappointed, because it would have been a boost to the area. They still believe that the Burren National Park needs a focal point that would stop visitors going onto private lands in the area

Source: www.rte.ie/news/2000/0306/burren.html

Its purpose is to act as a standard and guideline helping each signatory to develop high quality, sustainable tourism, defined as:

'any form of development or management of tourist activities ensuring the long-term protection and preservation of natural, cultural and social resources, and contributing in a positive and equitable manner to the economic growth and well-being of individuals living in, working in or visiting the protected areas'.

Figure 39
Tourist leaflet for the Aillwee Cave.

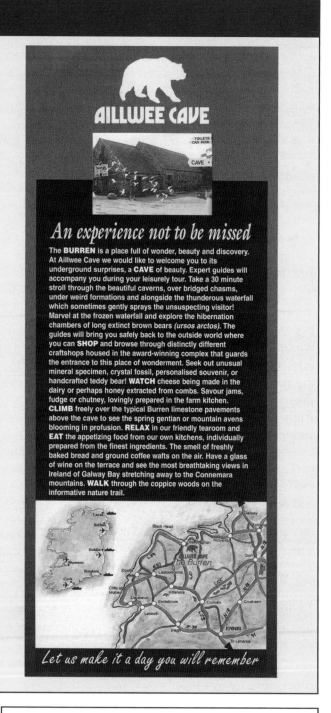

Tourism is a big draw to the Burren, but the balance between attracting the tourists to build up the economy in this rural area and keeping the environment in its natural state is a delicate one.

Tourists do bring in revenue:

■ The Burren captures 20% of Shannon's spend (£56 million)

■ Estimated additional 'day visits' – 0.4 million spending £4 million

The principles are:

- integrated approach towards tourism management;
- preservation of resources and reduction of waste;
- sharing the task of conservation and enhancement of the heritage;
- involvement of the local community;
- support to the local economy;
- development of an appropriate and quality tourism products;
- education and interpretation;
- sensitive marketing and promotion.

The World Conservation Union, in its report 'Parks for Life' stressed the fact that the pressures from tourism are growing rapidly and that numbers of tourists are expected to grow at an annual rate of 3–4.5% over the next 10 years. But, above all, protected areas are becoming increasingly popular as destinations.

In this context, careful and efficient planning, management and rigorous monitoring is needed in order to guarantee that tourism development in protected areas is successful and does indeed respect the natural, cultural and social environment of the area.

The Burren National Park contains a significant number of habitats and species (for example, turloughs and limestone pavements) which are required to be protected under the terms of the Council Directive 92/43/EEC of the 21st day of May, 1992, on the conservation of natural habitats and of wild fauna and flora.

More than 700 different flowering plants and ferns have been recorded in the Burren. Thus, although the Burren represents only 1% of the land mass of Ireland, 75% of the Irish native species are contained in the area, these includes alpine flowers like the Spring Gentian and Mountain Avens which can be found growing next to Mediterranean plants like the Dense Flowering Orchid.

Visitors are given education and advice:

- Help protect the wildflowers of the Burren: never pick flowers or remove plants or tamper with their habitats.
- Parking on the limestone pavement or grassland damages habitats.
- It is prohibited to pick or uproot plants in national parks and nature reserves.

activities

1. State fully one reason why it is important to conserve karst areas. [3]

2. (a) Name one area of limestone you have studied. [1]

   (b) State fully two causes of human pressure in your named area. [6]

   (c) Describe one management response to this human pressure. [3]

   (d) Name two groups of people that hold different views on how the area should be used. Describe and explain their views. [6]

HIGHER TIER

3. (a) With reference to a named area of karst, state fully two causes of human pressure in the area and explain two impacts this has had on the limestone. [12]

web link & extra resources

www.burrenbeo.com/

www.burrenpage.com/

www.iol.ie/~burrenag/

www.indigo.ie/~ironside/burren.htm

www.iol.ie/~burrenag/bm.htm

www.aillweecave.ie/ – an excellent overview of this cave system and features.

www.clarelibrary.ie/eolas/coclare/places/the_burren/burren_geology.htm

wwa.rte.ie/news/ – under this site search for issues around the Burren Centre, most of these were logged in 2000.

www.theburren.ie/ – looks at issues of tourism in the Burren.

www.ireland.com – has up-to-date information on issues relating to land use in the Burren.

Theme C Ecosystems and Sustainability

UNIT ONE
Distinct Ecosystems Develop in Response to Climate and Soils

Why do we see plants and animals that we do not see in Northern Ireland when we look at different parts of the world on television or when we go on holidays? Why would we think it unusual to find a pod of grazing rhinos along the Antrim Coast? Why is it that we generally see apples growing in the orchards of County Armagh rather than bananas? The reason is that certain plants and animals survive in areas because they are suited to the climate and soils of that area. This is what is called an **ecosystem** – a community of plants and animals within a physical environment. The physical environment of an ecosystem includes those elements that people have little or no influence over – the temperature, rainfall, soil, altitude and rock type. An ecosystem can vary in scale from an area of thousands of square kilometres, like a desert or a rainforest, to a small area of hedge or a pond.

Very large-scale ecosystems are called **biomes**. Biomes are huge areas with similar soils, plants and animals. These areas are related to the Earth's main climatic zones. Figure 1 below shows the location of the world's main biomes.

DISTRIBUTION OF ECOSYSTEMS

Tropical rainforest

The tropical rainforest is found in equatorial regions 5° north or south of the equator where average monthly temperatures are 25°C and annual rainfall is very high, often greater than 1800 mm. Areas where this biome is found include the Amazon Basin in South America, the Zaire Basin in Africa and parts of South-east Asia. The vegetation is the most luxuriant in the world made up of many layers, the tallest trees including mahogany and greenheart growing up to 50 m tall. The rainforest biome is also home to massive numbers of birds and insects.

Savanna

The Savanna biome is found in areas that have a hot and wet season with temperature over 26°C and a cooler dry season with temperatures around 21°C. The areas where this biome is found include sub-Saharan Africa, the Brazilian Plateau and northern Australia. Perhaps the best-known regions of this biome include Tanzania and Kenya, areas often associated with safaris and ecotourism. The vegetation includes grasses and

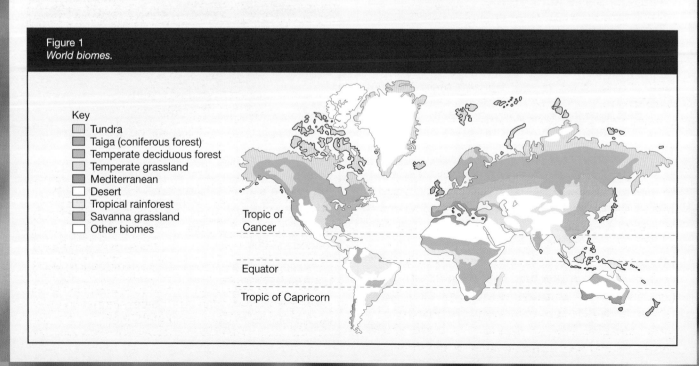

Figure 1
World biomes.

Key
Tundra
Taiga (coniferous forest)
Temperate deciduous forest
Temperate grassland
Mediterranean
Desert
Tropical rainforest
Savanna grassland
Other biomes

Tropic of Cancer

Equator

Tropic of Capricorn

drought-resistant trees such as the baobab. The animals include zebra, wildebeest, elephants and lions.

Desert

This is one of the world's most extreme biomes, with very little annual precipitation. The daily temperature range is very large, with daytime temperatures often in excess of 30°C and nighttime temperatures falling below 0°C. While the summers can be warmer than the winters, the area is characterised as being hot and dry all year. As a result some plants, referred to as succulents because they can store water in their tissues, have to survive long periods of drought. Others plants known as ephemerals complete their life cycle in a short period of time when there is available water. Cacti and thorn bush are typical of the region. There are limited habitats provided for animals, and any that can survive have to adapt to the hot and dry conditions. Examples include the camel, desert rat and various reptiles. Areas where this biome are found include the Atacama Desert in Chile, the Arizona Desert in the USA and the Great Australia Desert in Australia.

Mediterranean

The Mediterranean biome is typical of the Mediterranean region of southern Europe. This type of region can also be found in central California, central Chile and southern Australia. The climate is characterised by hot dry summers and warm wet winters. There is usually less than 700 mm of rainfall annually, and temperatures range from 12°C in the winter to 25°C in the summer. Similarly to the Savanna vegetation, the Mediterranean vegetation is drought resistant because it has to survive long periods without water in the summer months. The natural vegetation includes mixed forests of conifers and evergreens, as well as thorn and scrub. Humans have interfered with this vegetation significantly during thousands of years of cultivation, so it is not surprising to consider olive, almond and citrus trees as being most typical of the biome.

Taiga

This region is characterised by large areas of coniferous forests that stretch across North America and northern Europe. The climate is characterised by its extreme winters, with temperatures falling as low as –30°C in winter and rising to 15°C in summer. Because these regions are continental (inland) they do not experience large amounts of precipitation (around 400 mm annually). Any precipitation that falls in the winter does so as snow. The trees, fir, spruce and pine, have to adapt to these harsh winter conditions and then make the most of the short

summer growing season. Because there are limited varieties of plants, there are also limited varieties of animals. Depending on the location, elk, brown bear and wolves may be found in this type of biome.

One of the main differences between one biome and another is the biomass that is produced in the region. The **biomass** of an area is the total amount of living matter (vegetable and animal) within that ecosystem. Some areas such as tropical rainforests produce large quantities and varieties of living matter, whereas other regions such as deserts have very sparse vegetation cover.

One measure of the amount of vegetation produced in a region is Net Primary Productivity (NPP) which is measured in $g/m^2/year$ (dry weight). This allows the rate of growth to be compared in one region to be compared to another region. There are considerable variations from one region to another in terms of NPP, for example, tropical rainforests have a value of 2200 $g/m^2/year$, savanna regions have a value of 900 $g/m^2/year$ and desert regions around 90 $g/m^2/year$.

INTERACTION BETWEEN CLIMATE, VEGETATION AND SOILS

The reason why any ecosystem survives is because of the interaction that occurs between its climate, vegetation and soils. This interaction is outlined in Fig. 2. Natural ecosystems survive when there is balance between climate, soils and vegetation. Sometimes this balance can be upset when there is

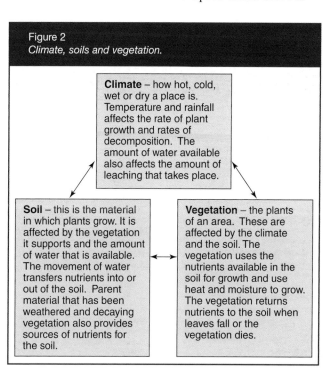

Figure 2
Climate, soils and vegetation.

Climate – how hot, cold, wet or dry a place is. Temperature and rainfall affects the rate of plant growth and rates of decomposition. The amount of water available also affects the amount of leaching that takes place.

Soil – this is the material in which plants grow. It is affected by the vegetation it supports and the amount of water that is available. The movement of water transfers nutrients into or out of the soil. Parent material that has been weathered and decaying vegetation also provides sources of nutrients for the soil.

Vegetation – the plants of an area. These are affected by the climate and the soil. The vegetation uses the nutrients available in the soil for growth and use heat and moisture to grow. The vegetation returns nutrients to the soil when leaves fall or the vegetation dies.

activities

1 State the meaning of the term **ecosystem**. [1]

2 What is meant by the term **biome**? [1]

3 What is the main difference between a biome and an ecosystem? [2]

4 Using the map below, name the type of biome found at places 1–10. [10]

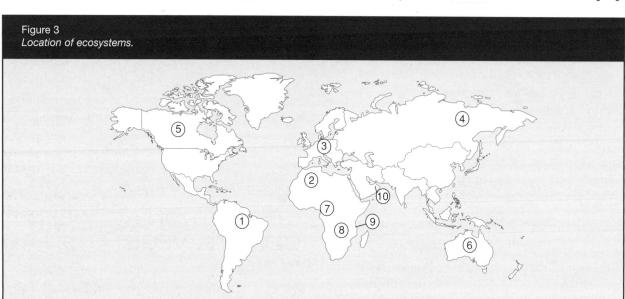

Figure 3
Location of ecosystems.

5 Complete the table below by matching the words and phrases listed below to each biome.

succulents baobab olive mahogany elk wildebeest camel   hot and wet all year

wet and dry season   little rainfall all year   very cold winters

	Climate	Vegetation	Animals
Tropical Rainforest			Monkeys
Savanna			
Desert			
Mediterranean	Summer drought		Goats
Taiga		Spruce	

6 For one of the biomes described above, design and draw an animal that would be able to survive in that biome. Your animal must be made up of parts of real animals. Make sure that the animal you design would be able to survive the climate and be able to eat. Describe how the animal would survive and give it a name. [10]

HIGHER TIER

7 State fully why the biomass for a tropical rainforest is much greater than that for a desert. [3]

activities

1 State the meaning of the terms soil, vegetation and climate.

2 Complete a copy of the diagram below by linking the following phrases, one has been completed for you.

- The weathered rock provides the soil with nutrients

- Dead and decaying vegetation returns nutrients to the soil

- The roots provide the plant with nutrients for growth

The climate provides heat and moisture for growth

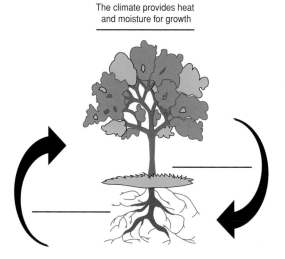

3 Think about the interrelationships or links that exist between the climate, soils, animals, plants and people in your local area – this could be your back garden, local park or local countryside.

(a) If the climate of your area dramatically changed by becoming wetter or hotter or drier, what would be the impact on the vegetation and the soils of the area? [1]

(b) If all the vegetation was removed from your area, what would be the impact on the soils and animals of the area? [1]

(c) How would the vegetation and soils of the area change if people no longer interfered with them? [1]

a dramatic change in one element. This happened when the Ice Ages affected Britain, the last of which ended around 10,000 years ago. Nowadays, the most significant challenges facing the world's ecosystems and biomes are the changes that humans have made. While there is evidence that humans may have begun to affect the climate, they certainly have dramatically changed the vegetation and the soils of many of the Earth's biomes.

Two further processes that operate in an ecosystem involve the **transfer of energy** from one part of the ecosystem to another and the **recycling of nutrients** within the ecosystem. This is when plants and animals use nutrients for growth and respiration and then recycle nutrients through waste products or death and decomposition.

PRODUCERS, CONSUMERS AND DECOMPOSERS IN A FOOD WEB

The energy that drives any ecosystem is obtained from the sun. The sun's energy allows plants to grow. The green plants that capture energy from the sun through the process of photosynthesis are the most essential elements of the **food chain**. These green plants are called the **primary producers**. It is important to realise that without photosynthesis there would be virtually no life on Earth.

Consumers obtain energy from these plants and pass it through the food chain. Consumers are the animals that either eat (consume) plants directly or else eat other consumers. There can be two or three levels of consumers in an ecosystem. At each level energy is used for growth and respiration, and heat is

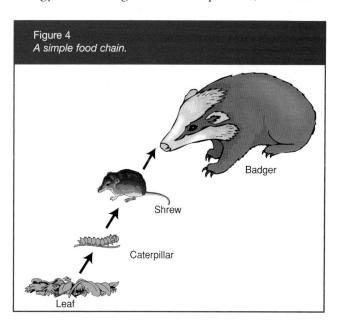

Figure 4
A simple food chain.

Badger

Shrew

Caterpillar

Leaf

also given off. As a result energy is said to be lost at each level through the food chain. This means fewer animals can be supported at each higher level. While there are fewer animals found at each higher level of the food chain they usually have a larger individual size than the previous level.

While the food chain shown in Fig. 4 is a simple example of the relationships that exist within an ecosystem, in reality the relationships tend to be more interdependent. As a result a food web can be a more realistic way of examining the relationships (see Fig. 5).

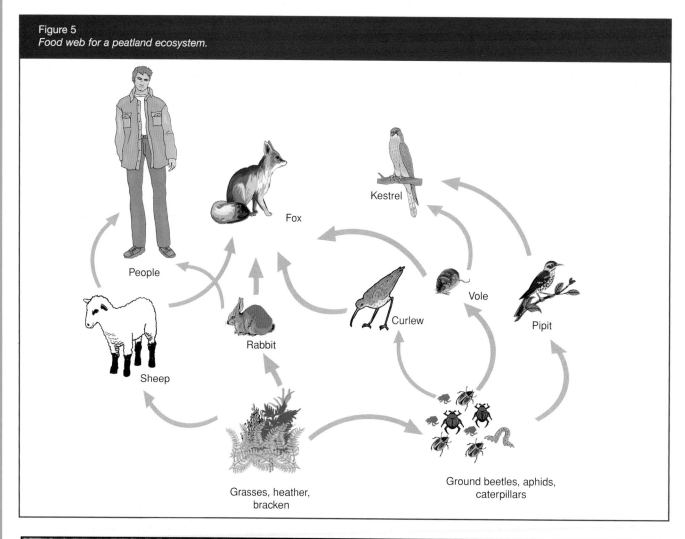

Figure 5
Food web for a peatland ecosystem.

People

Fox

Kestrel

Sheep

Rabbit

Curlew

Vole

Pipit

Grasses, heather, bracken

Ground beetles, aphids, caterpillars

activities

1 Name two producers and two consumers shown in the diagram above. [2]

2 Why are green plants known as primary producers? [1]

3 Complete a simple food chain with four links using the food web above. [2]

4 What happens to the size and number of each animal species moving up through the food web? [4]

HIGHER TIER

5 State fully the impacts on the food web if the numbers of rabbits were dramatically reduced because of disease. [3]

6 Describe the consequences for the plants and animals in the food web if there was a long summer drought. Explain your answer. [3]

Recycling of nutrients

The nutrients that are available within an ecosystem are recycled within the system. There are three main stores of nutrients – biomass, litter and soil. As previously stated **biomass** is the total amount of organic matter within the ecosystem – the animals and vegetation. Nutrients are transferred from the biomass when either plants or animals die, waste material is produced or leaves and branches fall from vegetation. This material then becomes part of the **litter**, the surface layer of vegetation. This litter layer breaks down to form **humus**, which is the decomposed organic matter in the soil which comes from plant and animal material. **Soil** is the organic (living) and inorganic (non-living) material in which the plants grow.

There are other inputs from this nutrient cycle including nutrients from rainfall into the soil and nutrients from broken down or weathered rock. **Leaching** is when nutrients are washed down through the soil. This usually happens when there is a lot of rainfall. Nutrients are also removed from the litter by surface run-off.

The transfer of nutrients within an ecosystem is shown by Fig. 6.

activities

Study Fig. 6.

1 Name the three areas where nutrients are stored.   [3]

2 Name two sources of nutrients for the ecosystems that are shown.   [2]

3 Describe two ways in which nutrients are lost from the ecosystem.   [4]

4 Describe the three ways nutrients are transferred between the nutrient stores.   [6]

5 What does the term leaching mean?   [5]

6 In what direction does leaching take place?   [1]

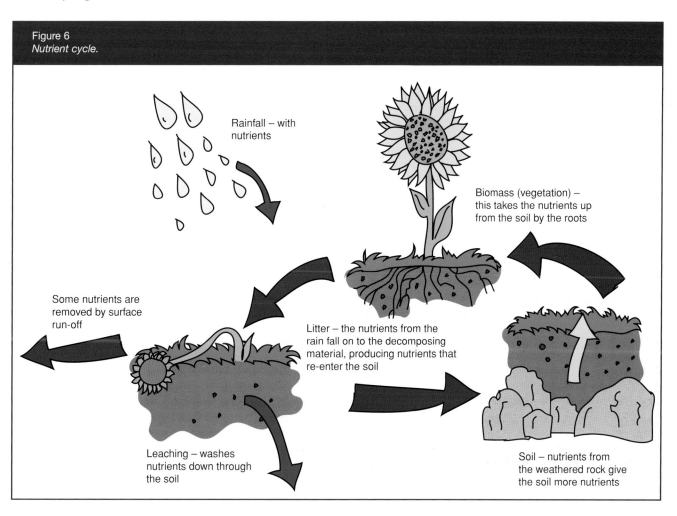

Figure 6
Nutrient cycle.

Rainfall – with nutrients

Biomass (vegetation) – this takes the nutrients up from the soil by the roots

Some nutrients are removed by surface run-off

Litter – the nutrients from the rain fall on to the decomposing material, producing nutrients that re-enter the soil

Leaching – washes nutrients down through the soil

Soil – nutrients from the weathered rock give the soil more nutrients

AN ECOSYSTEM IN NORTHERN IRELAND

Belvoir Forest is a large urban forest and conservation area on the outskirts of Belfast. It contains a variety of ecosystems including parkland, coniferous forest and deciduous forest. One area within Belvoir Forest known as The Big Wood Nature Reserve is predominantly a deciduous woodland ecosystem. Figure 7 below shows the location of Belvoir Forest.

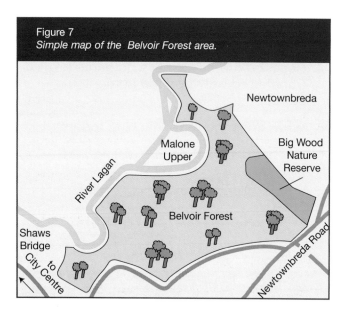

Figure 7
Simple map of the Belvoir Forest area.

Newtownbreda

Malone Upper

Big Wood Nature Reserve

River Lagan

Belvoir Forest

Shaws Bridge

to City Centre

Newtownbreda Road

Climate

The plants that grow and animals that survive in the Big Wood Nature Reserve in Belvoir Forest are able to thrive since they are suited to Northern Ireland's climate. Even though Northern Ireland is a relatively small area, it is important to realise that the climate varies from place to place. The highest annual rainfall is around 1600 mm in the Sperrins, the Mournes and the Antrim mountains. The driest places are the south and east shores of Lough Neagh and the east coast around Belfast, with approximately 800 mm of rainfall. The average annual temperature range is between 8.5 and 9.5°C at sea level, with July usually the warmest month. Northern Ireland's climate can be described as having mild winters, cool summers and rain throughout the year.

Soil

While the Big Wood Nature Reserve will mostly be viewed from ground level up, there are many important functions going on in the soils of this area. These can be considered by looking at a soil profile. It is important to know that the roots of the plants are reaching into the soil for nutrients and moisture

so the plants can obtain what they need for growth. These roots also hold the soil together and can help in the breakdown of the rock that lies underneath the soil through weathering. This provides the soil with nutrients.

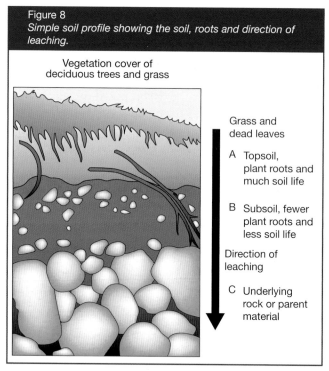

Figure 8
Simple soil profile showing the soil, roots and direction of leaching.

Vegetation cover of deciduous trees and grass

Grass and dead leaves

A Topsoil, plant roots and much soil life

B Subsoil, fewer plant roots and less soil life

Direction of leaching

C Underlying rock or parent material

As mentioned above one of the most valuable supplies of nutrients for the soil is the decaying plant and animal material (litter) that gathers on top of the soil. This material in the Big Wood Nature Reserve is broken down or decomposed by insects such as the Two Spot Long Horn Beetle, the Forest Bug, the White Legged Millipede and earthworm, as well as many different types of fungi and bacteria. Once this is decomposed, rainwater, which also contains nutrients, and the action of earthworms help move the material into the soil to form humus.

Soils are given names according to the conditions of the area that they were made under. Many soils in Northern Ireland are called Brown Forest Soils since they were formed under the deciduous woodland that covered much of Northern Ireland hundreds of years ago.

One of the most important factors to consider when looking at soils in Northern Ireland is **leaching**. Leaching occurs when nutrients are washed down through the soil and become difficult for the plants' roots to reach. This happens when there is a lot of rainfall in an area, or if the soil is on a slope. An example of a soil in Northern Ireland that is leached a lot and therefore loses a lot of nutrients is a Podsol.

Animals and vegetation

There are many plants that can be found in the Big Wood Nature Reserve in Belvoir Forest. In general there are tall mature trees, saplings, shrubs and ground-cover plants. Some examples of the plants found in the area include 200-year-old Sessile Oak trees, which support a wide variety of insect life, and the Ash tree, which is a very fast-growing native species. There used to be a lot of Laurel in this area, but this has been cleared recently since it was 'taking over' to a large extent and preventing the growth of other species. In clearings and around the edge of a stream that runs thorough the Nature Reserve, there are wild flowers such as bluebells and primroses. The snow-berry shrub is found around the bases of the more mature trees. While not strictly deciduous, other native species of trees found in this area include the Yew and the Scots Pine.

As a result of the conditions created by the different vegetation types, there is a wide variety of animals and birds in the Big Wood. While there are very few in evidence, the Red Squirrel can be found in this area. The badger is suited to the region as it is able to survive on earthworms, berries and other small mammals including the Field Mouse, which in turn feeds on berries and nuts. Birds in the area include the Blackbird, which eats insects and worms, the Robin, which feeds on insects and seeds, and the Long Eared Owl, which eats smaller mammals and birds.

(Original source:
www.bbc.co.uk/northernireland/schools/11_16/ks3geography)

activities

1 The plants and animals listed below can be found in the Big Wood Nature Reserve. Put each into the correct category in the table.

fox   bluebell   long horned beetle   maggot   hedgehog   rabbit   earthworm   primrose   field mouse   bacteria   nettles   wild garlic

Producers	Consumers	Decomposers
Sessile Oak	Robin	fungi

[6]

2 Name three sources of nutrients for the Big Wood Nature Reserve's soil.   [3]

3 What functions do the plant roots in this ecosystem serve?   [2]

HIGHER TIER

4 State fully the consequences for the Big Wood Nature Reserve if the Laurel Plant was allowed to grow in the area again?   [3]

5 Explain why some species in this ecosystem complete their life cycle earlier in the year than others.   [3]

CASE STUDY

TROPICAL RAINFOREST – THE AMAZON RAINFOREST IN BRAZIL

Figure 9
Aerial view of the Amazon.

One of the richest vegetated areas in the world is the tropical rainforest biome, which is found in equatorial regions. Nearly 40% of all the tropical rainforest left in the world is in the Brazilian Amazon. Although Brazil has lost approximately 58% of its forests, there is still over 1.8 million km^2 of tropical rainforest remaining which is among the largest amount of any country worldwide. However, approximately 2.3% of this forest is being cut down annually.

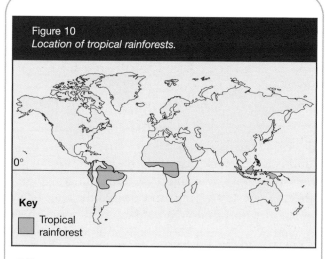

Figure 10
Location of tropical rainforests.

0°

Key

Tropical rainforest

Climate

The climate of the rainforest ecosystem is the driving force behind the vegetation and plants of the area. Consider what it would be like to have temperatures in excess of 25°C most days of the year, to have rain nearly every afternoon in heavy convectional storms, with over 1800 mm of rain throughout the year. These are the characteristics of the tropical rainforest climate.

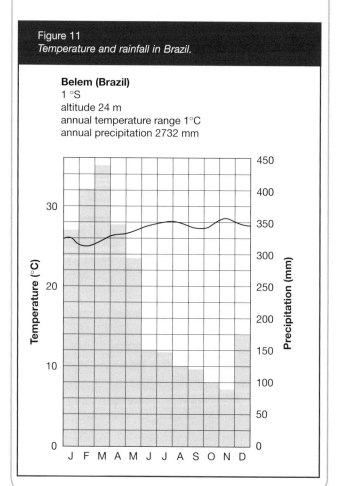

Figure 11
Temperature and rainfall in Brazil.

Belem (Brazil)
1 °S
altitude 24 m
annual temperature range 1°C
annual precipitation 2732 mm

continues

continues

The cause of the high temperatures is the location of the ecosystem between 5°N and 5°S of the equator. Here with the sun overhead temperatures are high every day. These high temperatures and the abundance of moisture means that there are convectional storms as the temperatures soar throughout the day.

Vegetation and animals

The tropical rainforest has an easily recognisable structure. With all the heat and moisture there is a massive amount of vegetation. The tropical rainforest is a biodiverse ecosystem. **Biodiversity** describes the number and range of plants and animals that are found within a particular ecosystem. A tropical rainforest with all its different habitats is much more biodiverse that a desert or polar ecosystem.

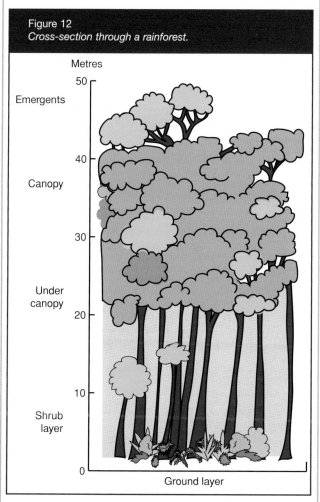

Figure 12
Cross-section through a rainforest.

Metres

Emergents — 50
Canopy — 40
Under canopy — 30, 20
Shrub layer — 10, 0
Ground layer

The Amazon rainforest, like others across the equatorial regions, is made up of five different layers. Working from the bottom up is the **ground**

layer. This is the forest floor which is dark and has little undergrowth. This area may flood for part of the year near to rivers. Only 1% of incoming light energy reaches the forest floor. A clearing may be found close to a river or if a large tree falls. This allows the seeds that have fallen to the forest floor to germinate, and struggle upwards towards the light.

The **shrub layer** reaches to about 5 m in height; here there are more plants such as ferns which can survive in the shadow of their larger relatives.

The **under canopy** consists of plants and trees up to 20 m tall. These are the young trees that are trying to make their way up to form part of the **canopy**.

The canopy is the massive covering of trees that stretch up to 40 m in height. Here there is most sunlight. These trees obtain 70% of the available light for the entire ecosystem. This is where trees such as mahogany, greenheart, rosewood and ebony are found. These hardwoods take a long time to grow and are a valuable resource for the people of the tropical rainforest regions.

Finally, there are the **emergents**. The trees at this layer tower above all the other trees as they stretch up to 50 m high. The emergents are able to obtain all the energy they need from the sunlight, and obtain their nutrients from the soil far below. These massive structures are supported by buttress roots which gives the trees stability. These trees are also branchless lower down so that all energy is concentrated in growing upward giving nutrients for the crown far above.

Many of the foods we eat today originally came from the rainforest. These include avocado, banana, black pepper, Brazilian nuts, cashews, chocolate/cocoa, cinnamon, cloves, coconut, coffee, cola, corn/maize, ginger, lemon, orange, pineapple, rice, sugar, tomato and vanilla.

Animals and insects

In the rainforest, there are many animals and insects, most of which never leave the trees but remain in the layer of the vegetation that best suits them. Toucans and parrots fly between the trees of the canopy and the emergents. Monkeys, sloths and butterflies have their homes among the branches of the canopy. Alligators and leopards are some of the top consumers within the ecosystem.

continues

continues

There are also mosquitoes, as well as hundreds of species of fish in the rivers and flooded regions of the forest.

A typical 6 km^2 patch of rainforest contains as many as 1500 species of flowering plants, 750 species of trees, 125 mammal species, 400 species of birds, 100 species of reptiles, 60 of amphibians, and 150 different species of butterflies.

Nutrients and soil

The tropical rainforest relies on the recycling of nutrients for its survival. Although the forest is luxuriant, the soil of the forest is quite poor. This is because the leaves that fall are quickly decomposed in the warm and humid conditions. As a result, the roots of the trees are quite shallow, and take up the nutrients before they get a chance to make their way from the humus into the soil. Nutrients that are moved into the soil are quickly leached downwards because of the heavy rainfall.

Once the trees are removed, and the supply of nutrients from the leaves is gone, the soils quickly lose their fertility. The trees having been removed also means that the soil is left open to **erosion** from the heavy rain. The rain carries huge amounts of loose soil away, as well as creating gullies. The increased amount of rain reaching the ground surface means that any remaining nutrients are quickly leached out of the soil.

One of the closest examples that can be found to a rainforest ecosystem in Northern Ireland is in the Palm House in the Botanic Park in Belfast. It is worthwhile to go there to experience the almost overbearing heat and to see some of the tropical plants that grow there.

activities

1 Name three places in the world where the tropical rainforest ecosystem is found. [3]

2 Describe the climate of the tropical rainforest. Remember that climate needs to be described using temperature and rainfall figures. Describe the climate over the whole year. [4]

3 Explain how convectional rainfall occurs. [4]

4 Draw a simple diagram of the structure of the tropical rainforest and label the following: [5]

Emergents   under canopy
shrub layer canopy   ground layer.

5 Why do so few plants grow on the forest floor? [3]

6 Why do the trees of the canopy and the emergents have branchless trunks? [2]

HIGHER TIER

7 State fully what is meant by a biodiverse ecosystem. [2]

8 Explain why the rainforest is considered to be a biodiverse ecosystem. [4]

9 Explain why the soils of the tropical rainforest is considered to be relatively infertile. [4]

10 Research and then explain with the help of a diagram how convectional rainfall occurs. [6]

UNIT TWO
Human Interference and Upsetting the Balance of Ecosystems

IMPACT OF HUMAN ACTIVITIES

Slash and burn, and large-scale deforestation in the Amazon

Figure 13
An Amerindian village.

The people that originally lived in the Amazon are called Amazonian Indians or Amerindians. There were estimated to be 6–9 million Amerindians living in the Brazilian rainforest in 1500. In 2000, less than 200,000 remained. Those that now remain have a reliance on the vegetation and the animals of the forest, as well as fish from the rivers of the area such as the mighty Amazon River. The main manner in which the Amerindians farm the land of the area is through a process of **shifting cultivation** or **slash and burn**. This means the people clear about a hectare of land with machetes and axes. They use some of the wood for fuel and building purposes. The rest is burnt, with the roots of some trees and an occasional taller tree remaining to protect the soil from the torrential rain. The ash from the burnt vegetation is ploughed back into the soil to provide nutrients for the crops to grow. Some of the crops cultivated include cassava, maize and manioc. After a period of 3 or 4 years the land loses some of its fertility since the crops have used up nutrients and the rain has also leached out the nutrients. So the tribe move on – they shift and start the whole process over again. This approach is considered to be **sustainable** because, although the people are using the forest, they are not exploiting it or overusing it. The area is being used, but the way it is being used ensures its survival for future generations. Once the area is left alone, the vegetation quickly starts to re-colonise the area. Within a period of 15–20 years, the area is almost as if it had not been touched. This is in stark contrast to large-scale deforestation.

Large-scale deforestation is at the other extreme of deforestation. Here the forest is cleared on a huge scale using chain saws, diggers and trucks. The reason why this occurs is to exploit the resources of the forest. Unfortunately, this is not a sustainable approach, since the area is often damaged beyond repair. One reason why this deforestation takes place is so that valuable trees such as mahogany and

Figure 14
Large-scale deforestation.

rosewood can be harvested. Sometimes many hectares of land are cleared to get at one tree! The forest is also cleared for roads. The Trans-Amazonian Highway was built to open up areas of the interior of the Amazon rainforest. Huge areas of the Amazon have also been cleared for cattle ranching and farming. Rubber plantations owned by companies such as Pirelli and Goodyear contributed to large-scale deforestation. The area has also been cleared to obtain minerals such as bauxite. An example of this is the large bauxite mine at Trombetas, north-east of Manaus. The creation of large-scale hydroelectric schemes such as the Tucurui Scheme on the Rio Tocantins resulted in the flooding of 4500 km^2 of forest. Finally, many people have been encouraged to move to the region from the overcrowded cities and shantytowns, with the promise of employment or a plot of land and farming equipment.

The Brazilian government is currently sponsoring a $45 billion Amazon development plan known as 'Advance Brazil.' This includes building new roads, river channelling and pipelines in the Amazon. When developments like this have occurred in the past there has been a massive increase in the number of people living in the region, moving from the countries overcrowded cities. Over the past 40 years, there has been a population explosion with numbers expanding from 2 million to 20 million people. As more people arrive, more land is cleared for agriculture and urban development. Several scientific studies predict that approximately 40% of Brazil's existing Amazon Basin forests will be lost or seriously damaged in the next three decades as a result of 'Advance Brazil'.

With all this deforestation the rainforest ecosystem is damaged more or less beyond repair. With the protective canopy of the forest removed the soil is open to the full force of the rain, resulting in the removal of massive amounts of soil and the creation of gullies. With the vegetation removed, the soil no longer has its supply of nutrients and as a result it quickly becomes infertile. People farming the area no longer have land on which they can produce crops, so they either move onto another area or migrate back to the cities that they have come from. Timber having been removed leaves vast areas of wasteland which cannot recover. Opencast mining techniques result in huge scars across the landscape.

Conservation in the Amazon rainforest

While there is much reporting of the negative impacts of deforestation in the Amazon, there are

activities

1   State fully the meaning of the term slash and burn.   [2]

2   State fully the meaning of large-scale deforestation.   [3]

3   Complete the table below to show the differences between the impacts of slash and burn and large-scale deforestation.   [16]

	Slash and burn	Large-scale deforestation
Size of area cleared		
Equipment used		
Who does the clearing		
Reason for clearing		
Impact on vegetation		
Impact on animals		
Consequences for the soil/land		
Sustainable or not?		

Figure 15
Protecting the Amazon.

In August 2002 'Tumucumaque National Park' in the Amazon was established creating the world's largest tropical forest protected area. The new park covers more than 3.8 million hectares and protects a significant part of the Amazon forest. Located in the Brazilian state of Amapá, the borders of the park were defined by WWF and the Brazilian environmental agency Ibama, in agreement with the Brazilian Ministry of the Environment. Its creation is a significant step towards fulfilling the pledge made by President Cardoso in 1998 to fully protect 10% or 41 million hectares of the Brazilian Amazon. This protection is needed to help secure the region's biodiversity.

Source: www.panda.com

also moves to protect the rainforest. **Conservation** is one means of protecting the rainforest and its people. Conservation may be defined as the protection of resources and landscapes for future generations. One way this is achieved is through a sustainable management approach of the ecosystem where the exploitation that takes place is not greater than the ability of the rainforest to replace itself. The examples below illustrate the scales at which conservation can take place. The conservation approaches are intended to protect the vegetation, soils and animals of the region as well as allowing the local communities to use the area in a sustainable manner.

Figure 16
Mahogany under threat.

There is currently a significant threat to the survival of the big leaf mahogany tree (*S. macrophylla*) in the Amazon due to illegal harvesting. Mahogany has been a prized wood for ship builders, furniture makers and woodworkers since the late 15th century. Since that time two of the three main mahogany species have become commercially extinct. The vast majority of the harvesting of this mahogany is illegal. Most of the wood ends up in the USA where the volume of the trade is huge. In 1998, for example, some 57,000 bigleaf mahogany trees worth over US $170 million were harvested and shipped to the US.

The current harvest of bigleaf mahogany is not sustainable since it is slow growing, taking 55–120 years to reach a commercially exploitable size. It often grows as isolated stands of even-aged trees, which are harvested by loggers. Up to 85 per cent of the adult population is often taken from each stand, significantly reducing the production of seed in subsequent years. While Brazil has had a near total ban on mahogany logging since 1996 national measures to control the illegal mahogany trade have so far failed. The illegal trade in mahogany undermines the biodiversity, and conservation status of the region's protected areas. The illegal trade has negative effects on Indian communities – some of which have had no previous contact with other people – whose land is home to mahogany reserves.

Protecting bigleaf mahogany would allow international commercial trade of the tree as long as the harvest is both legal and within sustainable levels. There is a international organisation with 160 member nations called the Convention on International Trade in Endangered Species of Wild Fauna and Flora (CITES) This organisation wants all governments to work cooperatively to ensure that the mahogany trade continues at sustainable and legal levels, in a well-managed manner that protects not only this species but the wider Amazon region and the indigenous people who live there.

Source: www.panda.com

Figure 17
Impact of conservation.

The "Manu Biosphere Reserve" in Peru is an area of land approximately 1.5 million hectares in area. It is an example of a protected area that is managed in a sustainable manner, where the resources of the area are able to be used, but are protected for future generations. The table below outlines the main impact this management has for the soils, animals, vegetation and the local community.

Manu Biosphere Reserve, Peru	Impacts of conservation
Soil	• maintaining vegetation cover protects soil from intense rainfall and erosion as well as overland flow • the nutrient cycle is maintained since the soil retains its litter and humus layer
Vegetation	• 1600 plants protected • large areas of forest are unlogged • natural regeneration encouraged • food trees maintained
Animals	• 1000 birds protected • 110 bats protected • top predators maintained
Local community	• the Indian population rights to live and work in the area are established • range of economic activities allowed including aquaculture, limited tree felling and agro forestry using fruit and rubber trees

activities

1 What is meant by the term conservation in relation to tropical rainforests? [3]

2 What are the main aims of establishing National Parks in the Amazon? [4]

3 Why does the Brazilian government believe it is important to conserve mahogany? [3]

4 What are the arguments for allowing mahogany to be harvested without protection? [4]

5 How can protecting mahogany help the soils of the rainforest? [3]

6 What is the role of CITES? [3]

HIGHER TIER

7 Why is it significant that there are 160 member countries of the organisation CITES? [3]

8 Why might protecting mahogany in one region lead to problems elsewhere? [3]

9 Outline using all three resources why conservation of the rainforest is important for the soil, vegetation, animals and the local community of the area. [12]

A PEATLAND ECOSYSTEM IN NORTHERN IRELAND – CUILCAGH MOUNTAIN, CO. FERMANAGH

An ecosystem in Northern Ireland that many people are familiar with is the peatland ecosystem. One of the reasons why people are familiar with peat is because it can and has been burnt as a source of fuel for generations. Peat is traditionally associated with the Irish landscape – the pile of turf outside a cabin, the smell of the peat fire burning or a donkey laden with creels of turf returning from the bog. These images of peat and peatland areas are traditional to Ireland because there have been large quantities of peat available for people in rural areas to use.

However, many generations of cutting has put pressure on the resource and, in areas where mechanised cutting has taken place, the peatland ecosystem has been altered beyond recognition.

Peat is developed when the plants in the bog die and build up in layers over thousands of years. Owing to the very wet conditions that exist, the plants decompose very slowly. It is these layers of slowly decomposing material that form peat. The peatland ecosystem is found in nearly every county of Northern Ireland, although in some areas it is quite inaccessible. There are two main types of peatland, blanket bogs and raised bogs. Blanket bogs are large areas of peatland that cover upland areas in the Antrim Plateau, the Sperrins and the Fermanagh uplands. Raised bogs are rounded mounds of peat that have developed over small lakes in low-lying areas. Here we are going to examine an area of blanket bog in Co. Fermanagh – Cuilcagh Mountain. The area lies some 20 km south-west of Enniskillen.

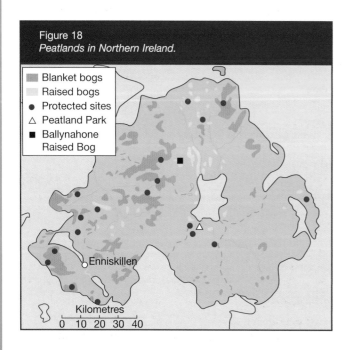

Figure 18
Peatlands in Northern Ireland.

Blanket bogs
Raised bogs
● Protected sites
△ Peatland Park
■ Ballynahone Raised Bog

Enniskillen

Kilometres
0 10 20 30 40

Figure 19
Traditional peat cutting.

Figure 20
Mechanised peat cutting.

Climate

Peatland areas are characterised as being very wet areas. The ground underfoot is soggy and boggy. This is because peatland areas experience large amounts of rainfall. Cuilcagh Mountain has up to 2000 mm of rainfall annually with rain falling throughout the year. Compare this to other parts of Northern Ireland such as the east coast that have only 800 mm annually. Upland areas are generally cooler than lowland areas. The effect of altitude on climate results in a rate of cooling of 1°C for every 100 m in height. Cuilcagh, at a height of 665 m, is on average 6.5°C cooler than sea level. The climate of this area is characterised by cool summers, cold winters and rain throughout the year.

Vegetation

In these conditions only certain plants survive. One of the most common species is heather. This species is well suited to the harsh climatic conditions, and grows extensively over the area. In addition to heather, there are sundews and butterworts which are plants that get extra nutrients by trapping insects on their leaves. There are also tough grasses, sedges, bog cotton, rushes and in the very wettest parts of the bogs sphagnum moss. Sphagnum moss can hold up to 20 times its own weight in water, which makes it an ideal plant to survive in the wet conditions of the bog. Most of these plants are slow growing and can survive even though the bog provides them with few nutrients. Most of the nutrients they obtain comes from the plentiful supply of rainwater.

Animals, insects and birds

The number and range of animals found on the bog is quite surprising. There are large quantities of insects such as dragonflies, butterflies, beetles, spiders, moths, flies, midges and pond skaters. The larger consumers include the Irish hare, foxes, badgers and stoats. There are many birds on the peatland including the skylark, meadow pipits, which provide a food supply for the hen harrier and merlin. The curlew, snipe and grouse can also be found in this area.

Impact of human activities – draining, peat extraction, grazing and afforestation in peatlands

There is increasing pressure to maintain and protect Northern Ireland's peatland ecosystem since the British Isles are home to most of Europe's remaining peatlands. Almost 90% of the peatlands in Northern Ireland have been damaged or destroyed in one way or another. David Bellamy, the well-known conservationist, has stated that the damage to the peatland ecosystem in Ireland is comparable to the damage done to the tropical rainforest ecosystem in South America, in terms of the impact humans have had on the soil, vegetation and animals of the ecosystem. One significant impact on peatland areas has been **afforestation**, which is the planting of trees. In Northern Ireland this is usually done with a uniform species of conifers over a large area which significantly alters the ecosystem of the area.

Some of the main human activities that have had a significant impact on the peatland ecosystem in the Cuilcagh Mountain Park area are outlined in the table below.

Threat	Nature of threat	Consequence
Overgrazing	Too many animals on the one area – usually sheep grazed.	Reduces the cover of heather and leads to the establishment of less productive grasses. Reduces habitats for other animals. Leads to areas of bare ground/ sheep tracks.
Drainage and reclamation	In the 1980s over 14 km of drains were dug on Cuilcagh to allow the bog to dry out before mechanised peat cutting was carried out.	Ground cover was reduced, water removed with plants and animals habitat significantly altered. Ground was no longer able to hold water and large-scale run-off occurred.
Mechanical peat extraction	The removal of peat on a large scale with the use of tractors and mechanised peat cutters. Farmers overusing their turbary rights.	Removes vegetation. Removes peat. Removes habitat for animals, insects and birds. Compacts peat layers. Causes lying water. Drying peat covers vegetation.
Afforestation	Planting of a single species of conifer over a large area.	Reduces the area of heather moorland. Restricts light to the forest floor reducing habitats for plants and animals. Alters the water balance – the trees' roots take up a lot of water. Needs the application of nutrients, usually artificial, which affects the nutrient balance of the surrounding area when the nutrients are leached out.
Supplementary feeding	Farmers give the animals (sheep) additional feed of silage, hay.	Leads to concentration of animals in one area and trampling. Grass seeds can germinate and spread over heather areas.
Uncontrolled burning	Areas burnt to control the growth of heather.	Forces animals to move or kills young/destroys nests/eggs. Destroys large areas of heather. Heather burning can be used positively to encourage regeneration – uncontrolled burning is the threat to the peatland ecosystem.

activities

1 Name three producers and three consumers in the peatland ecosystem. [6]

2 What conditions are needed for the formation of peat? [4]

3 Why is sphagnum moss suited to the peatland ecosystem? [3]

4 In what way have sundews and butterworts adapted themselves to survive in the peatland ecosystem? [4]

HIGHER TIER

5 Why do conservationists believe that the peatland ecosystem is an important ecosystem to conserve? [3]

ICT

6 Using the information from the table below produce a climate graph (temperature/rainfall graph) for Belfast, Northern Ireland. [8]

Belfast	J	F	M	A	M	J	J	A	S	O	N	D
Average monthly temperature (°C)	4	4	6	8	11	13	15	15	13	10	7	5
Average monthly rainfall (mm)	80	52	50	48	52	68	94	77	80	83	72	90

activities

1 Study the table on page 82. For any three threats to the peatland ecosystem outlined, explain why people may have found it necessary to carry out these activities. [6]

2 The threats listed above had a negative on the Cuilcagh Mountain ecosystem. Use the table below to outline the impact on the soils, vegetation, animals and local community of any **three** of the threats described. [12]

Threat	Impact on soil	Impact on vegetation	Impact on animals	Impact on local community
1.				
2				
3.				

Conservation and benefits to soil, vegetation, animals and the local communities

As has been seen throughout the 1980s and early 1990s human activities have had a significant impact on the soil, vegetation, animals and local community in the Cuilcagh area. The interference by humans with one aspect of the ecosystem had major consequences for the rest of the ecosystem. For example, the mechanised cutting of peat on Cuilcagh was affecting the rate of water flowing into the Marble Arch Caves several miles away. This in turn was affecting the amount of tourists that were able to visit the caves, which affected the amount of income that was being generated in the area from tourism.

It became clear that something needed to be done to try to halt and, if possible, reverse the impact of some of the impacts that humans had had on the ecosystem. Therefore by 1998 Fermanagh District Council set up the Cuilcagh Mountain Park with the aims of protecting the intact areas of the blanket bog of Cuilcagh Mountain, restoring damaged areas, and making people more aware of the environmental value and rarity of peatland habitats and wildlife. This involved the **conservation** of the area and meant that new management practices had to be put in place. Conservation is when an area is protected

and improved upon so that it is protected for future use. The **management** of an area such as the Cuilcagh Mountain Park involves the day-to-day and longer term running of the area by interested groups and individuals so that the area can be used to its full potential. Fermanagh District Council have the most important management role in the running of the Cuilcagh Mountain Park. Some of the most

Figure 21
Repairing damage to peatland.

significant management and conservation approaches adopted are outlined below.

- Cuilcagh Mountain Park is within the West Fermanagh and Erne Lakeland Environmentally Sensitive Area (ESA) Scheme. This is an agri-environmental scheme operated by the Department of Agriculture NI to promote less intensive methods of farming and promotes good conservation practices.

- The Cuilcagh Mountain Park also lies within ASSI (Area of Special Scientific Interest) as declared by the Environment and Heritage Service because of the important peatland habitats, geology, geomorphology, and flora and fauna of the area.

- DUCHAS (the Heritage and Wildlife Service of the Republic of Ireland) along with the British and Irish governments have proposed that the Cuilcagh blanket bog will be protected as a cross-border Special Area of Conservation (cSAC).

- Cuilcagh Mountain Park is also a 'Ramsar' site, which identifies it as a wetland of international importance adding further protection for the plants and animals of the area.

- The Department of Agriculture and Rural Development for Northern Ireland (DARD) are working with sheep farmers on Cuilcagh to encourage good conservation practices. In addition the Forestry Service of DARD own a large amount of the bog and have decided to conserve it instead of planting coniferous forest, turning the area into a nature reserve where bog plants and animals are protected

- Fermanagh District Council have combined the visitors' centre for the Marble Arch Caves with an interpretative centre for Cuilcagh Mountain Park. This provides information, education services, exhibitions and a café, all of which provide employment for up to 50 people.

- Environment and Heritage Service have introduced positive measures to protect the blanket bog.

- Royal Society for the Protection of Birds provides information and protection for the many birds that have their habitat in the area.

- The Marble Arch Caves and Cuilcagh Mountain Park are jointly the first European Geopark in the United Kingdom which will benefit the area both in scientific terms and economic development.

Some of the impacts that these conservation and management approaches have had on the soils, vegetation, animals and local communities are outlined in the table opposite.

activities

ICT

1 Imagine you are the Environment Officer for Fermanagh District Council. You have to give a presentation report to the council outlining how Cuilcagh Mountain Park is being conserved. In your report you must discuss the following topics:

 (a) The threats that Cuilcagh Mountain Park faced in the past;

 (b) The reasons why Cuilcagh Mountain Park needed to be protected;

 (c) The types of management and conservation strategies that are carried out in the area;

 (d) The benefits the conservation and management strategies bring to the soil, vegetation, animals and the local communities of the area.

Your report may be produced using Microsoft PowerPoint.

[20]

	Impact of conservation and management
Soil	Building small dams to retain water to keep the peat in its more natural soggy state.
	Prevention of the wholesale cutting of peat keeps area of peatland intact.
	Prohibiting mechanised cutting protects the peat and prevents it becoming compacted.
Vegetation	Fewer sheep by lowering stocking density prevents overgrazing of the heather.
	Good burning practices prevent the destruction of large areas of vegetation.
	Protection status such as ASSI, ESA and Ramsar make damage to vegetation illegal.
	Less planting of coniferous forest keeps the natural vegetation intact.
Animals	ASSI and ESA status provides protection for the animals of the peatland.
	The protected vegetation maintains the habitat and food source for the animals.
	RSPB provides protection for birds.
	Restrictions on burning protect the young animals at vulnerable times of the year.
Local communities	Jobs provided in the interpretative centre.
	Protecting the peat helps control the discharge of the rivers flowing from Cuilcagh Mountain into the Marble Arch Caves. This helps secure the tourist attraction and therefore protects jobs.
	Increased tourism in the Fermanagh area provides employment opportunities in restaurants, B&Bs and the service industry generally.
	The economy of the area is strengthened with increased tourism.

UNIT THREE
Management of Ecosystems and Sustainable Development

ADVANTAGES AND DISADVANTAGES OF ECOTOURISM IN KENYA

Ecotourism is regarded as an environmentally friendly type of tourism where people tour natural habitats in a manner that is ecologically friendly. The International Ecotourism Society (TIES) defines ecotourism as 'responsible travel to natural areas that conserves the environment and sustains the well-being of local people'. This can be contrasted with conventional tourism to places like the Costa del Sol in Spain. This 'package holiday' type tourism is considered to have a more negative impact on aspects of the local environment and the way of life of the local people.

Ecotourism in places like Kenya is big business, bringing many advantages to its local communities, vegetation, animals and soils. Tourism is very important for the economy of the country. However, a balance has to be reached between encouraging tourism on one hand, and allowing too much tourism, on the other. If areas have too many tourists – even if they claim to be ecotourists – they can have a negative impact on the people, soil, vegetation and animals of an area.

Background – about Kenya

Kenya is an independent republic in East Africa. It is bordered by Sudan, Ethiopia, the Somali Republic, Tanzania and Uganda. The country has many outstanding features including Mount Kenya, the second highest mountain in Africa after Kilimanjaro, the Rift Valley, Lake Turkana and an area of Lake Victoria.

Tourism

Tourism is Kenya's most important foreign exchange earner attracting hundreds of thousands of visitors every year with numbers constantly increasing. Tourists are attracted to Kenya mainly by the country's national parks and game reserves, where wildlife can be seen and photographed in its natural surroundings. These national parks are wildlife sanctuaries that are intended to protect the animals in them.

Most of Kenya's wildlife has been preserved in national parks and game reserves in a natural undisturbed state. A total of 8% of Kenya's land area is designated as national parks. The most important of these parks and reserves can be seen in Fig. 22 and are briefly described below.

Within the national parks qualified staff are needed to guide the tourists, drive the minibuses and work in hotels. This provides jobs for local people who may otherwise have to survive by pastoral farming or growing crops. However, most of the jobs are only available in the peak season, in summer, and the people are often unemployed for the rest of the year.

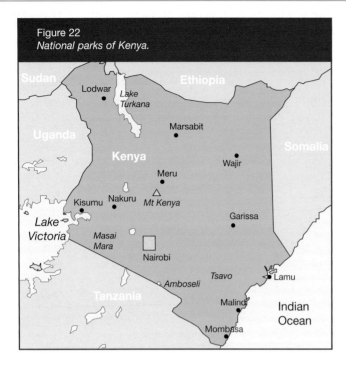

Figure 22
National parks of Kenya.

Tourists can have a huge impact on the ecosystem they go to see. They are driven around the national parks in specially adapted minibuses with semi-open tops so that they can safely view and photograph the animals. There are regulations forbidding buses to go closer than 25 metres to any animal, but these rules are hardly ever obeyed and because of this the animals can be disturbed. The safari buses are only allowed on designated roads but this also can be ignored by bus drivers attempting to get as close as possible to animals so the tourists can take good photographs. As new tracks are worn soil erosion increases, causing significant erosion in areas that were previously vegetated. Tours can also be taken in hot air balloons, which can affect the animals. The gas burners used by the balloons make a very loud noise, and the balloons cast huge shadows over the ground, which disturbs and frightens wildlife.

Accommodation is needed for the large numbers of tourists within the national parks. Sometimes purpose-built lodges are erected which can provide jobs for local people during construction. Local people can also get jobs cooking, cleaning and providing other services for the tourists. The lodges mean that more people are concentrated in the one area and this can put a strain on water supplies, waste disposal and energy supplies. Roads, which are built for the benefit of tourists, can be used by the local people which can improve access and attract investment.

The tourists who come to the national parks can sometimes be less than sensitive to the needs of the local communities. For example, tour buses regularly visit the Maasai villages, the homes of the traditional inhabitants of the Maasai Mara Park. Tourists buy souvenirs that the Maasai have produced, but they can treat the villages as tourist attractions rather than authentic settlements where people live and work. The impact of more Western influences can have a negative impact on the traditions in the area.

Sometimes Maasai tribesmen offer their services at campsites as guards. There are other benefits for the Maasai such as schools and a water supply funded by the park's profits. However, most of the revenue made from the parks goes to the tour companies and the government, not to the people who originally were the owners and guardians of the area.

The following passages and images are taken from advertisements for safaris to Kenya. They give an outline of the type holidays that can be enjoyed by ecotourists on safari.

Tsavo National Park
Area: 21,288 km², Alt: 2250–2440 m

The largest park in Kenya is known for its diverse and striking landscapes as well as its roving wildlife, especially the large animals such as Elephant, Rhino, Hippos, Eland and Antelope. Tsavo is also known for the fierce man-eating lions of the 1900s. Vegetation is mainly arid savannah, while the region is generally hot and dry most of the year.

Amboseli National Park
Area: 3810 km², Alt: 1150 m

Famous for the stunning scenery, with the 19,450 feet Mount Kilimanjaro dominating the view, this park is also known for its wildlife, especially the prolific elephant herds. The vegetation is varied but dominated by savannah which allows for viewing of the plains game, ranging from lions, leopards and giraffe.

Maasai Mara Game Reserve
Area: 1800 km², Alt: 1500–2200 m

Probably the most famous wildlife reserve in the world, the 'Mara', as it is popularly known, is the best place to see a range of game, from the big cats to the grazers and river game. The Mara also hosts the spectacular and primal wildebeest migration, which occurs annually as the wildebeest and other animals migrate south across the Mara River.

Figure 23
Images of Kenya.

Figure 24
Maasai Mara Safari.

Maasai Mara National Reserve

Main Features

- Annual Migration
- Exceptional Game Viewing: Lion, Leopard, Cheetah, Elephant
- Mara River Hippo Pools
- Traditional Maasai Culture
- Views across the Plains and Rolling Hills

The Maasai Mara National Reserve is probably the most famous and most visited Reserve in Kenya. It offers breathtaking views (as seen in the film 'Out of Africa', much of which was filmed here), an extraordinary density of animals including 'the Big Five' (lions, leopard, elephant, buffalo, rhinoceros) and many varieties of plains game.

An impressive feature is the annual migration of wildebeests, zebras and gazelles from the plains of the Serengeti that cross the Tanzanian border and rivers to reach the Mara's grasslands from late June, tracked by predators: lion, leopard, cheetah, and hyena, and circled by vultures as their journey unfolds. Their dramatic river crossings are a reality for tourists visiting in early July–August.

Apart from the seasonal migration, game viewing is excellent year round. Game includes elephant, black rhino, buffalo, plains zebra, hartebeest and big cats. The rivers are home to hippo and crocodiles. 452 species, 53 of which are raptors.

Accommodation ranges from stone built lodges to luxury tented camps. The area to the North owned by the Maasai offers great game-viewing, game walks and night games. Safari operators set up private camps for small groups seeking exclusive and traditional safaris out of the Reserve. In the Reserve are four tented camps and two lodges.

Original source: www.shoortravel.com

Figure 25
Safari in Kenya.

Safari in Kenya

This safari exhibits the wild and stunning spectacle of Africa, with a game viewing extravaganza in the Maasai Mara and Amboseli National Parks, the latter set against the breathtaking back drop of Mt Kilimanjaro, Africa's highest mountain. The safari includes a one-night stay at the Lake Nakuru Park, which is world famous for its rare species of pink flamingos.

Perfect for the first-time visitor, this safari gives just a taste of the huge variety on offer, whetting the palate for a return trip. Try a little camping in Amboseli; with your own vehicle, driver, naturalist guide and camp cook, you will be very comfortable in your large, igloo style tents with mattresses, bucket showers and traditional safari long drop toilets. This is a true camping experience!

Follow up your camping adventure with comfortable, excellent quality lodges with friendly, helpful staff, ensuite facilities and stunning views for the remainder of your safari. Once again, you'll have your own personal vehicle and driver/guide to escort you through the wilderness, ensuring you a wealth of information about

the animals, plants, birds and peoples you meet along the way.

Perfect on its own, or as an extension to an East African business trip, this trip offers a little bit of everything!

Original source: www.shoortravel.com

activities

1 What is meant by the term ecotourism?
[2]

2 Name five animals that could be seen on safari in Kenya. [5]

3 What sort of accommodation would an ecotourist stay in while on safari in Kenya? [3]

4 State fully why you think National Parks were set up in Kenya. [4]

5 Ecotourism can bring benefits to an area. Using Kenya as an example, complete a copy of the table below outlining some of these benefits to the soil, animals, vegetation and local communities. [8]

HIGHER TIER

ICT

Ecotourism can also bring disadvantages to a region. Working with a partner, imagine you are advisors to the Tourism Minister in Kenya. Draw up a set of proposals stating how you are going to ensure that the national parks and the soil, vegetation, animals and local communities in them are going to be developed for tourism in a sustainable manner. [10]

Benefits ecotourism brings		
Soil		
Vegetation		
Animals		
Local Communities		

Theme D Population and Resources

UNIT ONE
Distribution and Density

PHYSICAL AND ECONOMIC FACTORS AFFECTING DISTRIBUTION AND DENSITY

People live all over the world's surface. There are areas with lots of people and areas where only a few people live. **Population distribution** is the way in which people are spread out over an area with some areas crowded and other areas more sparsely populated. The world's population distribution is said to be uneven. There are only a very few areas that have an even population distribution and these have generally been planned, like parts of the Canadian Prairies or the Dutch Polders.

There are many reasons why people decide to live in one area or another. Sometimes these decisions have been made generations before by family members who settled in the area for reasons that now seem difficult to understand. Some areas have many obvious reasons why people would live there, such as a pleasant climate, beautiful scenery, job opportunities or good farming land.

There needs to be a way that the numbers of people living in different areas can be compared. **Population density** measures the number of people in a given area, usually a km^2. It is found by dividing the population of a country or region by its land area. For example, Ireland has a population density of 53 persons per km^2, Spain 78 persons per km^2, the

Figure 1
Population density by country.

People per km^2

- <10
- 11–25
- 26–100
- 101–200
- >201

United Kingdom a population density of 243 persons per km^2, Japan 334 persons per km^2 and Bangladesh, one of the most crowded places on the Earth, has a population density of 953 persons per km^2!

Population density can be shown on a choropleth map. Choropleth maps show the population density of a region by using colours or shading, the darker the colour or the denser the shading the higher the population density.

Figure 1 shows the world's population density by country for 2002.

In general, choropleth maps showing population are useful because they give an idea of the overall population density for the country and allow comparisons to be made. However, such maps only show average values. For example, the USA is made up of very densely populated areas such as New York and Los Angeles as well as very sparsely populated states such as Alaska. These differences in the population density of the US are not evident from the population density map. There can also be sharp divisions at political boundaries. For example, the Republic of Ireland has a much lower population density than Northern Ireland since Northern Ireland shares its population density figures with the rest of the United Kingdom.

There are physical and economic factors that affect population distribution and density. **Physical** factors are natural or environmental factors such as relief, drainage, climate, soil and natural vegetation. **Economic** factors are related to money, making a living or wealth. Figure 2 shows how physical and economic factors affect the population distribution and density in six places across the globe.

activities

1 State the meaning of the term population density. [2]

2 How is population density calculated? [2]

3 What is the meaning of the term population distribution? [2]

4 Complete a copy of the table below by giving one physical and one economic factor which affect the population density of each of the six areas shown in Fig. 2. [10]

Place	Physical factor	Economic factor
Antarctica		
Sahara		
Himalayas		
Bangladesh		
Tokyo		
Mexico		

HIGHER TIER

5 State fully the main advantages and disadvantages of choropleth maps for showing population distribution. [4]

6 Using an atlas, research other ways in which population distribution and density can be shown. What are the advantages and disadvantages of these approaches? [6]

ICT

7 Using information from these pages produce a bar graph showing the different population densities of a range of countries. Include a title on your graph and describe what the graph shows about world population density and distribution. [10]

Figure 2
Physical and economic factors affecting population density.

Sahara – like other deserts this has high temperatures reaching up to 40°C in the summer with a daily temperature range often over 50°C because of cold nights. The main difficulty in desert regions is the lack of water, with less than 250 mm of rainfall per annum. In these arid areas the lack of vegetation, limited supplies of water, extremes of climate and limited communications means that the only people who survive in this region are nomadic pastoralists.

Bangladesh – With 953 people per km², the country is one of the most crowded on earth. Over 80% of the population live in rural areas with many of these being subsistence farmers. These people are able to survive since the rich soils and the monsoon climate allow for three rice harvests a year. However, the increasing population is forcing more people to live in the lowest lands that are most likely to flood or be affected by tidal waves. This flooding can lead to the destruction of crops and loss of lives.

Himalayas – this huge range of more than 30 mountain peaks is the highest mountain region in the world. The highest mountain in the range, Mount Everest, is the world's highest mountain at 8848 m. While there are many people living in the foothills of these mountains, the upper slopes are virtually uninhabited, with extremes of climate and near impossible communications. 2002 was designated the Year of the Mountains because of their importance in providing water to people living in lowland regions.

Mexico – Mexico City and its surrounding metropolitan area has a population of over 18 million people making it the second largest urban areas in the world. Mexico City is the capital city, and has grown at a rapid rate due to in-migration from the surrounding rural areas and high birth rates. People want to live in Mexico City because of the job opportunities. More than half of Mexico's industrial output is produced in or near Mexico City. Major highways and railways radiate from the city to all parts of the country. A large international airport is located east of the city. Colonias (new settlements) have been developed on the edges of the city, and now rural migrants move there, with fewer going into slums in the city centre.

Antarctica – the most sparsely populated areas in the world. No permanent residents, only camps of research scientists and occasionally tourists. It is extremely cold with temperatures lower than – 50°C in the winter. Resources are strictly controlled by international agreement. It is very inaccessible. The first person born in Antarctica was Emilio Palma, the son of the commander of Argentina's Esperanza Base, on 7 January 1978.

Tokyo – is Japan's largest city and capital. It is one of the most densely populated cities in the world with up to 13 000 people per km². The city of Tokyo has a population of approximately 8 million people and the Tokyo metropolitan area has a population of almost 26 million making it the largest urban area in the world. Tokyo is the country's financial, industrial, commercial and cultural centre. The city's industries are concentrated along the shores of Tokyo Bay and include heavy industry, light manufacturing and hi-tech products, as well as a wide variety of consumer goods.

Figure 3
Images of Spain: from crowded beaches to barren rural landscape

Distribution and density of population in Spain

Spain is a country that most people in Northern Ireland are familiar with because they think of it as a tourist destination. While tourism is very important when considering the population density and distribution of Spain, there are easily identified physical and economic factors that affect Spain's overall population density and distribution.

Spain makes up the 80% of the Iberian Peninsula, with Portugal making up the remaining 20%. The population of Spain is close to 40 million people, with the majority being urban – 75% of the population live in towns and cities, and 25% live in rural areas. The population density of Spain is approximately 78 people per km^2.

There are two main cities in Spain – Madrid and Barcelona. Madrid is the capital and is found almost in the centre of the country. It has a population of approximately 3 million people. Barcelona, which was originally a natural harbour, is found on the Mediterranean coast and has a population of 1.8 million. The main reasons for so many people living in these two cities are the same reasons that attract people to cities the world over. Madrid is the capital and therefore the centre of government. Madrid is also an important industrial, financial, administrative and service centre. It is one of the country's three richest regions along with the Basque country in the north-east and Catalonia, the region around Barcelona. The

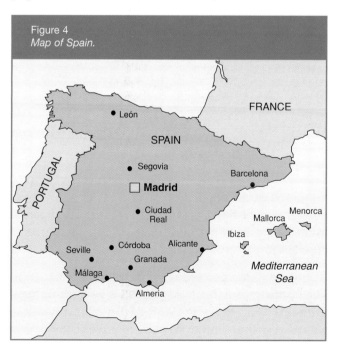

Figure 4
Map of Spain.

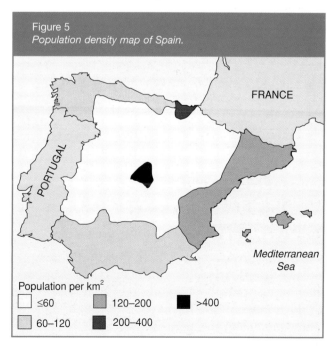

Figure 5
Population density map of Spain.

Population per km^2

- ≤60
- 60–120
- 120–200
- 200–400
- >400

richer 'half' of Spain lies to the east, and the poorer regions lie to the west.

While Madrid is the capital, the second-largest city, chief port and commercial centre is Barcelona, the capital of Barcelona province and the Catalonia region. Barcelona also has a thriving tourist industry. Other important cities include Valencia, with approximately 800,000 people, a manufacturing and rail centre; Seville with 700,000 people, capital of Seville province and the Andalusia region, a cultural centre; Saragossa with 600,000 people, capital of Saragossa province and the Aragón region, another industrial centre; and Bilbao with 370,000, a busy port. The economic activities of these regions keep population densities high.

The climate of Spain has a significant impact on the population densities and distribution. The south and east coasts are ideal for tourists; especially those from northern Europe looking for almost guaranteed sunshine and high temperatures in the summertime. As a result, not only are there huge numbers of tourists, but also there are large numbers of people employed in the tourist industry keeping population densities high along the coast.

Lower population densities are found in regions such as the Maseta, which is the central plain of Spain. This region has a continental climate, with hot dry summers and cold dry winters. Here the population density is low with the main activity being extensive agriculture.

The mountainous regions of Spain also have much lower population densities: the Pyrenees, which form the boundary with France, are inaccessible and have climatic extremes. This is also true for the Cantabrian Mountains along the northern coast.

Population densities are also lower in Galicia in the north-west. While this is an important tourist and cultural region, there is still a significant proportion of the population employed in fishing and agriculture. Incomes here are among the lowest in Spain and there are significant levels of out-migration.

Spain's population density is uneven. There are many different reasons why people live in the different areas of the country, and, while there are only some examples given here to represent the overall trend, it can easily be seen how physical and economic factors influence the distribution of the population.

activities

1 Name two densely populated regions of Spain. [2]

2 Name two sparsely populated regions of Spain. [2]

3 Produce two spider diagrams – one diagram for economic factors and one for physical factors that affect Spain's population density and distribution. You could use symbols to illustrate the factors and remember to add specific place names. [8]

4 Draw a map of Spain labelling the places mentioned in the section above. Add short notes to the map to describe and explain Spain's population distribution and density. [10]

UNIT TWO
Population Changes Over Time

CHANGES IN BIRTHS, DEATHS AND SIZE OF WORLD POPULATION SINCE 1700

Population growth occurs when **birth rates** are greater than **death rates**. Birth rate is a measure of the number of babies born in a region. It is expressed as the number of live births per thousand of the population in an area per year. Death rate is a

measure of the number of people dying in a region and is expressed as the number of deaths per thousand of the population of an area per year. The reason why birth and death rates are expressed per thousand and not per hundred (per cent) is so that whole numbers can be used.

Population grows according to the difference between the birth rate and the death rate. This is called **natural increase**. The rate of the world's natural increase has varied over time, and from place to place.

For thousands of years the population of the world was very low, growing to only 500 million by around 1650. However, from 1700 onwards the population started to grow quite rapidly and then exploded around the beginning of the 20th century, as can be seen in Fig. 6. For the last 50 years, world population has multiplied more rapidly than ever before, and more rapidly than it will ever grow in the future.

The world's population in 1980 was about 4.4 billion people (one billion is one thousand million). The 6 billion mark was reached in 1999 and it is now estimated that the world's population will grow by 50%, from around 6.1 billion today to 9.3 billion by 2050. The rate of population growth has varied over time and from place to place. In 1800, the vast majority of the world's population (86%) lived in Asia and Europe, with 65% in Asia alone. However, from around 1950 the population of the world's LEDCs began to increase dramatically. The proportion of the world's population living in LEDCs is expected to continue to rise. Most growth will occur in the world's 48 least-developed countries which will nearly triple in size, from 658 million to 1.8 billion people.

Population geographers (demographers), statisticians and politicians have examined this rapid population growth for decades. Governments of the world are concerned that if a very rapid rate of growth continues then certain areas will become overpopulated. **Overpopulation** is when there are too many people in an area for the resources available to support an adequate standard of life. There are areas in the world, for example the Sahel region of Northern Africa, where people suffer from famine, lack of water and where the land is being overcultivated. These conditions are an indication of overpopulation. In MEDCs overpopulation can be indicated by traffic congestion, pollution and high land prices. Since overpopulation is related to

population and resources some people argue that the problem of overpopulation is not simply about the population growing too quickly, but is related to the way the world's resources are distributed between richer and poorer countries. Overpopulation is discussed more fully later in this chapter.

There are different estimates of how quickly the world's population will grow, based on current growth rates and projected growth rates. Figure 7 shows three different estimates of what the world's population could be by 2050.

Population growth rate is a measurement of the rate at which the population is increasing. It is obtained by dividing the natural increase by 10 to get a percentage. The 2000 population growth rate of 1.4%, when applied to the world's 6.1 billion population, gives an annual increase of about 85 million people. This is shown by the medium projection on Fig. 7. Because of the large and increasing population size in the world, the number of people added to the global population will remain high for several decades, even as growth rates decline.

To help show how quickly population can grow consider a country with a population of 1 million that has a population growth rate of 3% per year. It will add 30,000 persons the first year, almost 31,000 the second year, and so on. Therefore at a 3% growth rate it will have doubled its population in around 23 years. Geographers consider a population's **doubling**

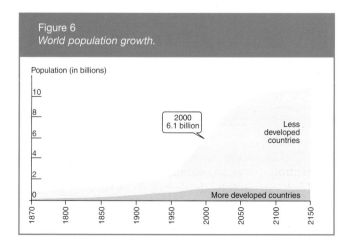

Figure 6
World population growth.

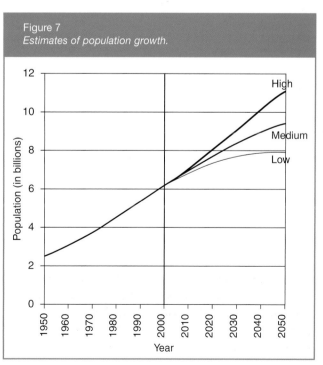

Figure 7
Estimates of population growth.

time when comparing rates of population growth from one country to another. The doubling time for a population can be roughly determined by dividing the current growth rate into the number '69' (69 ÷ 3 = 23 years).

Growth rates of 2% in many LEDCs means populations will double in around 35 years. However, as birth rates start to fall, the rate of population growth may decline. The more economically developed countries of North America and western Europe have growth rates of less than 1%, meaning that their proportion of the world's population is expected to stabilise. However, in some eastern European countries, death rates are higher than birth rates meaning a negative growth or a population decline, for example Russia (–0.6%), Estonia (–0.5%), Hungary (–0.4%) and Ukraine (–0.4%).

activities

1 What is meant by the terms birth rate, death rate and natural increase? [6]

2 There are many factors affecting the birth rates and death rates of a region. Some of these factors are listed below. Indicate in the boxes whether the factor mainly affects birth rates or death rates and whether it would cause an increase or decrease. [17]

Factor	Birth rate(✓)	Increase/ decrease	Death rate(✓)	Increase/ decrease
Clean water supply	❑	....	❑	....
Artificial methods of contraception	❑	....	❑	....
Public Health Acts of Parliament	❑	....	❑	....
Vaccinations against infectious diseases	❑	....	❑	....
Improved nutrition	❑	....	❑	....
Later marriage	❑	....	❑	....
Children needed for work	❑	....	❑	....
High levels of infant deaths	❑	....	❑	....
Large elderly population	❑	....	❑	....
Improved hospital care	❑	....	❑	....
Sewage system	❑	....	❑	....
Compulsory education	❑	....	❑	....
Rural population	❑	....	❑	....
Urban population	❑	....	❑	....
Short life expectancy	❑	....	❑	....
Desire for more material possessions	❑	....	❑	....
Emancipation of women	❑	....	❑	....

activities

3 Complete the following table which has information about birth rates, death rates, natural increase and the population growth rate for a range of countries. [9]

Country	Birth rate (per thousand)	Death rate (per thousand)	Natural increase (per thousand)	Population growth rate (per cent)
India	26	9		1.7
UK	12			0.1
Italy	9		–1	
Kenya		14	18	
France		9	3	

4 Which countries have a growth rate greater than 1%? [2]

5 Which countries have a growth rate of less than 1%? [2]

6 According to the table, which countries would you consider to be MEDCs and which would you consider to be LEDCs? Give reasons for your answer. [6]

7 What problems might a country face that has a negative growth rate? [4]

HIGHER TIER

8 Use the formula outlined earlier to calculate the **population doubling time** for each of the countries in the table. [5]

9 Outline some of the consequences for a country that has population doubling time of 35 years? [6]

ICT

10 Enter the information from the table in question 3 into a spreadsheet. Which AutoSum commands will you use to quickly calculate natural increase and population growth rate? [4]

11 Produce a line graph using the information from your spreadsheet to compare the birth rate, death rate and natural increase of the countries shown. [4]

12 Describe and explain what the graph shows about the birth rates, death rates and natural increase of the countries shown on your graph. [8]

13 Study Fig. 6. What was the world's population in 1750, 1900 and 2000? [3]

14 Explain why it is often stated that the world's population exploded from the beginning of the 20th century. Use figures to illustrate your answer. [4]

15 Why do you think population growth is occurring most rapidly in LEDCs and least in developed countries? [4]

HIGHER TIER ICT

16 Study Fig. 7. What is the difference between the highest and the lowest estimates of population growth by 2050? [1]

17 There are three projections of population growth shown in Fig. 7. Why do you think estimates of population growth vary so widely? [3]

THE DEMOGRAPHIC TRANSITION MODEL

The rate at which the population of countries change is a result of natural increase – the difference between birth rates and death rates. This can be shown using a graph called the **Demographic Transition Model**.

The Demographic Transition Model (DTM) or **Population Change Model** shows birth rates, death rates and the total population of a country over time. It can also illustrate countries stages of development according to their birth rates and death rates. Stage 1 shows a limited stage of development and Stage 4 is characteristic of a MEDC.

The DTM can be compared to filling a bath with water (see Fig. 8). The water flowing from the taps can represent the birth rate. The water leaving the plughole shows the death rate. The water level in the bath shows the total population. The rate at which the water levels rise depends on how quickly the water either goes into or comes out of the bath.

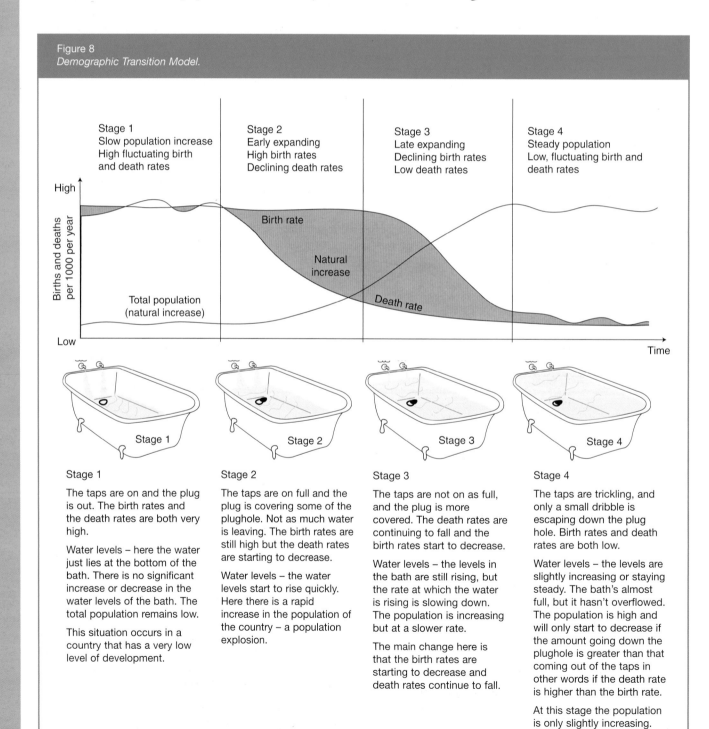

Figure 8
Demographic Transition Model.

Stage 1
Slow population increase
High fluctuating birth and death rates

Stage 2
Early expanding
High birth rates
Declining death rates

Stage 3
Late expanding
Declining birth rates
Low death rates

Stage 4
Steady population
Low, fluctuating birth and death rates

Stage 1

The taps are on and the plug is out. The birth rates and the death rates are both very high.

Water levels – here the water just lies at the bottom of the bath. There is no significant increase or decrease in the water levels of the bath. The total population remains low.

This situation occurs in a country that has a very low level of development.

Stage 2

The taps are on full and the plug is covering some of the plughole. Not as much water is leaving. The birth rates are still high but the death rates are starting to decrease.

Water levels – the water levels start to rise quickly. Here there is a rapid increase in the population of the country – a population explosion.

Stage 3

The taps are not on as full, and the plug is more covered. The death rates are continuing to fall and the birth rates start to decrease.

Water levels – the levels in the bath are still rising, but the rate at which the water is rising is slowing down. The population is increasing but at a slower rate.

The main change here is that the birth rates are starting to decrease and death rates continue to fall.

Stage 4

The taps are trickling, and only a small dribble is escaping down the plug hole. Birth rates and death rates are both low.

Water levels – the levels are slightly increasing or staying steady. The bath's almost full, but it hasn't overflowed. The population is high and will only start to decrease if the amount going down the plughole is greater than that coming out of the taps in other words if the death rate is higher than the birth rate.

At this stage the population is only slightly increasing.

Figure 9
Stages of the Demographic Transition Model.

	Birth rates	Death rates	Countries
Stage 1	The birth rates are high to ensure that some babies survive into adulthood, to work the land and look after their parents. As a result while many babies are born, many die and life expectancy is low.	Death rate is high since there is little access to medicines, and low levels of sanitation and hygiene leave people liable to infectious diseases. Poor food supplies and dirty water also causes higher death rates.	This stage is typical of native tribes people such as the Amerindians of the Amazon rainforest.
Stage 2	Lots of babies are born but fewer are dying therefore the population is increasing. There are early marriages and the tradition of having large families. With religious observance families have many children. The levels of education are relatively low and so people are not fully aware of methods of family planning. There are still many infants dying so parents continue to have large families in case some do not survive.	The overall decrease in death rates comes about due to improved nutrition and a wider availability of clean water. There are improvements and increased levels of medical care with vaccinations available for infectious diseases for example cholera and dysentery. An improvement in living standards, and safer living and working conditions and improved diet has a significant impact. There is a rise in life expectancy.	Countries such as Afghanistan, Somalia and Burkina Faso – the world's Least Developed Countries can be found in Stage 2.
Stage 3	The birth rates start to fall because of an increase in equal opportunities for women. There is greater access to education and an increase in female employment. Marriages are later and women are increasingly interested in career development rather than having children. There is an increase in the availability of artificial methods of contraception which further controls the number of children in a family. The government may also promote family planning measures encouraging parents to have fewer children to slow population growth.	Increased medical care, an awareness of health and welfare issues, the provision of a health service and pension provision all increase life expectancy and decrease the death rate. This is further helped by a continuing improved diet and access to clean drinking water for most of the population as wealth increases and public services are improved.	Mexico, Egypt and Colombia are examples of countries that can be found in Stage 3 as their birth rates continue to fall and their death rates remain low.
Stage 4	Parents have a small number of children, on average 2–3 per family. There are high levels of female employment and education, and widespread availability of family planning. The government may be concerned about the decreasing rate of growth of the population, and may give financial bonuses to those having larger families.	The death rate remains low as there are high levels of medical provision. There is an effective government that protects the health and welfare interests of the population and a long life expectancy	Many western European countries are found at this stage, the United Kingdom, France and Spain, as well as the United States.

In the same way the total population in a country rises or falls according to the differences in the birth and death rates.

POPULATION STRUCTURE

The DTM is one way in which the population of a country or region can be illustrated. Another way is a population pyramid. A population pyramid shows two main population characteristics: the age and sex of the people living in the area, this is also known as the **composition** or structure of the population. A population pyramid can also be called an age–sex pyramid.

The pyramid is divided in two: the left-hand side shows the male population and the right-hand side the female. The bars that make up the pyramid represent the age of the population, divided into five year groups or cohorts: 0–4, 5–9, 10–14 and so on. The numbers along the bottom of the pyramid represent the percentage of the total population in each category. Sometimes population pyramids show total numbers instead of percentages. Percentages are generally preferred since they allow for easy comparisons to be made between countries.

Figure 10 is a population pyramid for the United Kingdom for 2003. The points labelled

activities

1 Which three elements of population does the Demographic Transition Model show? [3]

2 Complete the table below by adding the terms high, low, increasing, decreasing and steady in the correct places. [16]

	Stage 1	Stage 2	Stage 3	Stage 4
Birth rate				
Death rate				
Natural increase				
Total population				

3 Name two countries or regions that could be placed in each stage. [2]

4 Complete the graph by adding your own birth rate, death rate and natural increase according to the descriptions above. Remember to add a title, key and label the axis. [8]

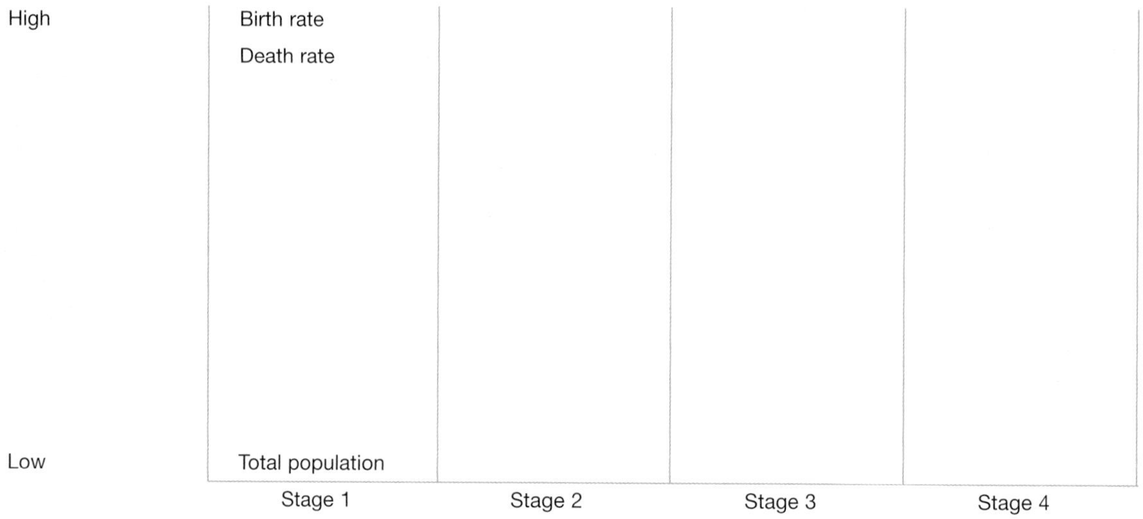

High

Birth rate
Death rate

Low

Total population

Stage 1   Stage 2   Stage 3   Stage 4

HIGHER TIER

5 Could there be a further stage added to the Demographic Transition Model? [1]

6 Describe any further stage of the Demographic Transition Model in terms of birth rates, death rates and total population? [3]

7 What countries that you have read about previously in this chapter would fit this stage? [2]

explain how to interpret what the pyramid is showing.

Figure 11 shows how population pyramids can be compared to different stages of the DTM. If a country has a high birth rate (DTM Stage 1 or 2) the base of the pyramid tends to be wide. If it has a low birth rate (DTM Stage 3 or 4) the base is much narrower. When there is a high death rate (DTM Stage 1 into Stage 2) then the pyramid narrows quickly, but when the death rate is low (DTM Stages 3 and 4) then the sides tend to be more even. A country that has a long life expectancy has a much taller pyramid (Stage 3 or 4), but if it has a shorter life expectancy then the pyramid is much shorter (DTM Stage 1 or 2).

In general:

■ Less developed countries have a wider pyramid showing a higher birth rates, rapidly decreasing sides representing a lot of infant deaths (high infant mortality) and a high death rate, and they are quite compact with few elderly, representing a shorter life expectancy.

■ More developed countries have a narrow base indicating lower birth rate, they have relatively even sides, showing low levels of infant mortality, a lower death rate and good health care, and they tend to be tall with a large proportion of people in the older age groups showing a long life expectancy.

Figure 10
UK population pyramid, 2003.

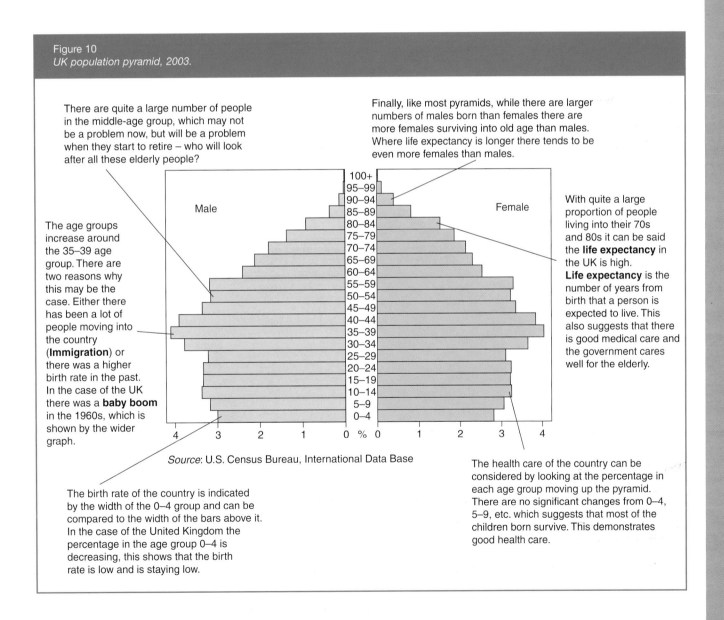

There are quite a large number of people in the middle-age group, which may not be a problem now, but will be a problem when they start to retire – who will look after all these elderly people?

Finally, like most pyramids, while there are larger numbers of males born than females there are more females surviving into old age than males. Where life expectancy is longer there tends to be even more females than males.

The age groups increase around the 35–39 age group. There are two reasons why this may be the case. Either there has been a lot of people moving into the country (**Immigration**) or there was a higher birth rate in the past. In the case of the UK there was a **baby boom** in the 1960s, which is shown by the wider graph.

With quite a large proportion of people living into their 70s and 80s it can be said the **life expectancy** in the UK is high. **Life expectancy** is the number of years from birth that a person is expected to live. This also suggests that there is good medical care and the government cares well for the elderly.

The birth rate of the country is indicated by the width of the 0–4 group and can be compared to the width of the bars above it. In the case of the United Kingdom the percentage in the age group 0–4 is decreasing, this shows that the birth rate is low and is staying low.

Source: U.S. Census Bureau, International Data Base

The health care of the country can be considered by looking at the percentage in each age group moving up the pyramid. There are no significant changes from 0–4, 5–9, etc. which suggests that most of the children born survive. This demonstrates good health care.

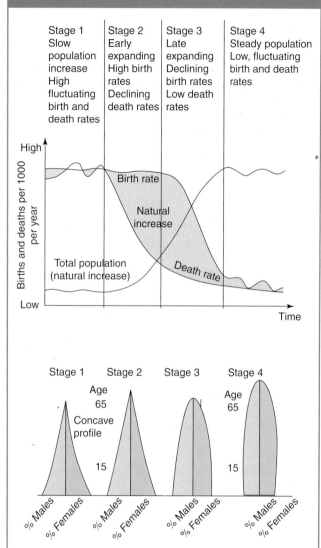

Figure 11
The Demographic Transition Model and population pyramids.

Stage 1 Slow population increase High fluctuating birth and death rates	Stage 2 Early expanding High birth rates Declining death rates	Stage 3 Late expanding Declining birth rates Low death rates	Stage 4 Steady population Low, fluctuating birth and death rates

Ireland 2003

Age	Male %	Female %
0–4	3.7	3.4
5–9	3.6	3.3
10–14	3.7	3.4
15–19	3.8	3.7
20–24	4.2	3.9
25–29	4.1	4.0
30–34	3.9	3.9
35–39	3.5	3.5
40–44	3.4	3.5
45–49	3.2	3.2
50–54	3.1	3.0
55–59	2.7	2.7
60–64	2.1	2.0
65–69	1.8	1.8
70–74	1.4	1.5
75–79	0.9	1.4
80–84	0.6	0.9
85–89	0.3	0.9
Total	**50**	**50**

2 Figure 12 is a population pyramid for Afghanistan in Asia, which is one of the world's least developed countries. Using Fig. 10 as a guide, complete six labels around a copy of the pyramid to show the characteristics of the population. [6]

Figure 12
Population pyramid for Afghanistan, 2003.

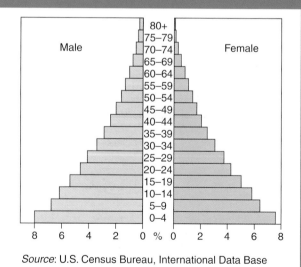

Source: U.S. Census Bureau, International Data Base

3 Describe the main differences between the shape of the pyramid for the UK (a MEDC) and the pyramid for Afghanistan (a LEDC). [4]

1 Complete a population pyramid for Ireland in 2003 using the information in the table opposite. Remember to put the males on the left-hand side and the females on the right-hand side with the age groups in the centre of the pyramid. Give your graph a title. [8]

activities

1 Add the following labels to a copy of the population pyramids shown in Fig. 13 below. [8]

High birth rate   Low death rate   Long life expectancy
Short life expectancy   Large elderly population
Large youth population   Low birth rate
High death rate

Figure 13
Two population pyramids from the Demographic Transition Model.

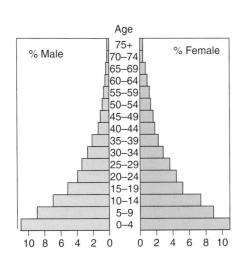

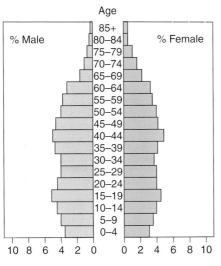

2 Which pyramid is more characteristic of a MEDC and which is more characteristic of a LEDC? Explain your answer fully. [6]

Migration

The population of any country changes as a result of birth and death rates. However, migration of people into or out of a country can also have a significant impact on the population of a country. **Migration** is when people move into or out of a region on a permanent basis. **Immigration** or in-migration is when people move into a country or region. **Emigration** or out-migration is when people exit from or leave a country or region. Figure 14 illustrates that the population of an area is a balance between birth rates, death rates and migration.

Figure 14
Factors leading to population change.

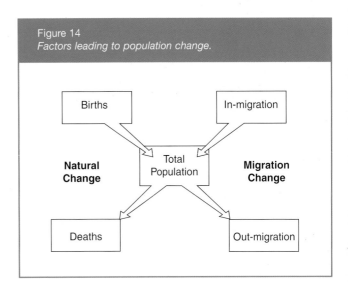

There are a range of factors that lead people to move from one region to another, as will be outlined later in this chapter. The impact of migration can often be seen in a country's population pyramid. The following two examples show the impact of immigration on the United Arab Emirates, and the consequences of emigration on Albania. As population geographers it is important to recognise the signs that show migration has taken place as well the consequences this migration has on the country. Notice that the pyramids for these examples have been constructed using total numbers rather than percentages. This gives an indication of the actual numbers of migrants into and out of the regions.

Population pyramids and migration

Out-migration – Albania

Albania is one of the poorest countries in Europe. This former communist state relies heavily on foreign aid and has a lot of debt. As a result of poverty and the need to escape from political upheaval there has been significant out-migration

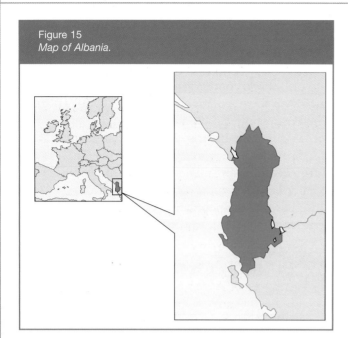

Figure 15
Map of Albania.

- The loss of population has an impact on service provision.

- Often the most healthy and active members of the population leave.

- Since it is mainly males who migrate, it could be assumed that in future years they will return home to their families.

In-migration – United Arab Emirates

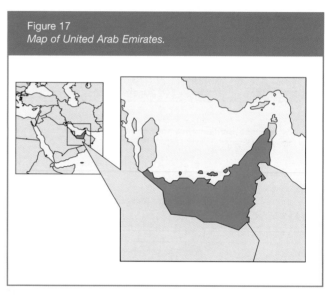

Figure 17
Map of United Arab Emirates.

from the country with 300,000–400,000 Albanians working outside the country and sending money back. These remittances help support the families left at home, but the migration has had a significant impact on the population structure of the country.

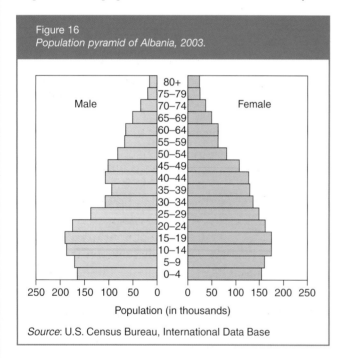

Figure 16
Population pyramid of Albania, 2003.

Source: U.S. Census Bureau, International Data Base

- There is a significant indent in the pyramid in the 25–39 age group for males. This emphasises that males are generally more migratory than females.

- The migrants are economic migrants.

- The younger population is more migratory than the older population.

- There is a male to female imbalance in the country.

The United Arab Emirates (UAE) is a federation of seven independent states lying along the east central coast of the Arabian Peninsula. It currently has a population of less than 2 million. The population grew significantly during the 1970s and the 1980s as a result of in-migration. It is estimated that today only 25% of the population are nationals originally from the UAE. Migrant workers have come from India, Pakistan, Bangladesh and Iran, as well as other countries. Migration has occurred because of the significant employment opportunities not only in the petroleum industry, but also in natural gas exploration, aluminium production and construction, as well as education and nursing.

- There is significant in-migration, especially of males.

- There has been an increase in the birth rate in the recent past as a result of the increasing population.

- The migrants are economic migrants.

- The pyramid is significantly unbalanced reflecting an unbalanced male to female ratio in the country.

- Money is being taken out of the economy of the country.

- The mix of different cultures creates tensions between different groups.

Figure 18
Population pyramid of United Arab Emirates, 2003

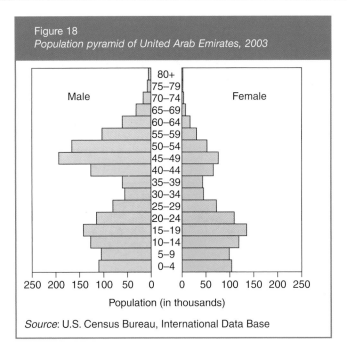

Population (in thousands)

Source: U.S. Census Bureau, International Data Base

- There is evidence of decreasing numbers of males in the older age groups suggesting migrants may have returned home on retirement.

activities

Answer the following questions using the United Arab Emirates and Albania as examples.

1 What is meant by the term migration? [2]

2 State fully the difference between emigration and immigration. [2]

3 What are the reasons for so many migrants leaving Albania? [3]

4 Outline some of the reasons that attracted migrants to the United Arab Emirates. [6]

5 State fully the problems that face a country such as Albania that has lost a large proportion of its population due to migration. [6]

6 Outline some of the benefits that migration may bring to a region of out-migration. [6]

7 What problems may an area face as a consequence of large-scale in-migration? [6]

Dependency

Figure 19 on the next page shows two population pyramids, one for the UK (a MEDC) and one for Afghanistan (a LEDC). The pyramids clearly show that there are different proportions of the population of the two countries in different age groups. Population geographers are most interested in three general age groups: the youth dependent age group 0–14, the aged dependent age group 65+ and the independent or active population 15–64. In general terms, the dependent population of any country has to be supported by the independent or working population. One way the level of dependency is examined is through the dependency ratio.

The dependency ratio of a country is calculated by dividing the dependent population by the active or independent population and multiplying by 100. This is shown by the following formula:

$$\text{Dependency ratio} = \frac{\text{youth dependent (0–14)} + \text{aged dependent (65+)}}{\text{working population (15–64)}} \times 100$$

In general terms, MEDCs have a dependency ratio of between 50 and 75, but because of the increasing numbers of elderly in MEDCs this is growing. LEDCs often have a dependency ratio greater than 100 because of the large number of children in the population.

Youth dependent population

- The **youth dependent population** is the youngest group of the population. They are usually dependent on parents or guardians for food, clothing and shelter. They are also dependent on the independent population for education and health care which is provided by the government from the taxes of the working population.

- This population tends to be much larger in LEDCs than in MEDCs due to higher birth rates and larger proportions of young people in the population. However, in many LEDCs, as birth rates are falling the proportion in the youth dependent population is falling also. Countries with a large youth dependent population have to provide a lot in terms of education and medical provision.

Independent or working population

- The **independent** or **working population** are those aged from 15 to 64. It seems strange for us in the British Isles to consider 15 year olds to be in the working population, but in other societies where there is much less in terms of education, starting work at this young age is quite normal.

■ The government taxes the working population, and from the taxes governments pay for health care, education and social services such as child benefit or unemployment benefit. The working population also makes pension contributions where the funds accumulated are used to support the pensions of the aged dependent population.

■ The amount that is taken from the working population in taxes and pension contributions varies from country to country. The difference between what is provided by a country such as the United Kingdom is hugely different from what is

activities

Look at the two population pyramids for Afghanistan and the UK shown in Fig. 19 below.

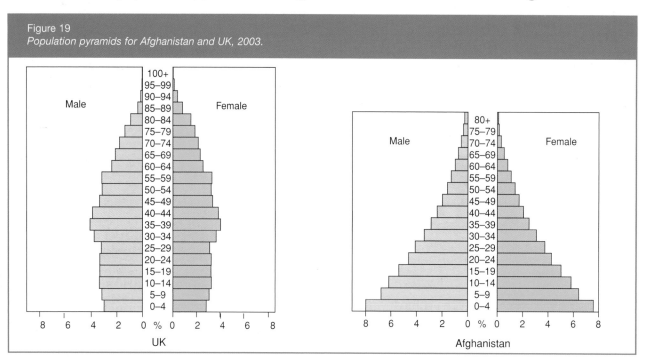

Figure 19
Population pyramids for Afghanistan and UK, 2003.

1 Which country has the largest youth dependent population? [1]

2 Which country has the largest aged dependent population? [1]

3 What are the implications for the population structure of each country in terms of:

■ Education provision;

■ Health care;

■ Employment;

■ Social services. [8]

HIGHER TIER

4 Using the figures shown on the pyramids, calculate the dependency ratio for each country. Notice that in these examples the population pyramids have been constructed using total numbers rather than percentages. [4]

5 What do the figures suggest about the level of development of the country? [6]

provided in a LEDC such as Somalia. In LEDCs where there is a large working population there may not be enough jobs to go around. This leads to problems of unemployment, underemployment and an informal economy.

Aged dependent population

- The **aged dependent members** depends on the working population for pension contributions as well as health care and other social service benefits.

- In MEDCs with an increasing elderly population, there is an increasing burden on the working population to pay for health care and pension provision. As people have a longer life expectancy then pensions have to be paid for longer periods of time. Some governments such as the UK's are encouraging people to work beyond 65 so they can support themselves for longer.

- While an elderly population is not a major problem for LEDCs at present, a growing proportion of older people in the future will have to be planned for. A previously high birth rate and improved health care will increase numbers in a LEDC's aged dependent group through time.

Population change and structure in Italy and Mexico

Population change in Italy

Figure 20
Map of Italy.

Italy is a country in the south of Europe that has undergone quite a transformation in its population structure over the past 30 years. It has gone from being a country that has traditionally been characterised by large families, to one where there is a danger of the population beginning to decline. This is due to birth rates being lower than death rates. There is, as a result, a large and increasing elderly population who have to be supported by an ever-decreasing working population.

The graph below shows that while Italy's population is still growing, the rate of growth is very small and beginning to go into decline.

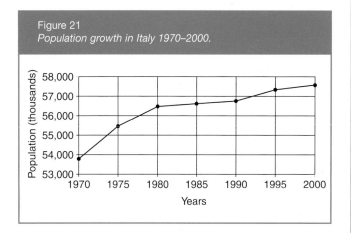

Figure 21
Population growth in Italy 1970–2000.

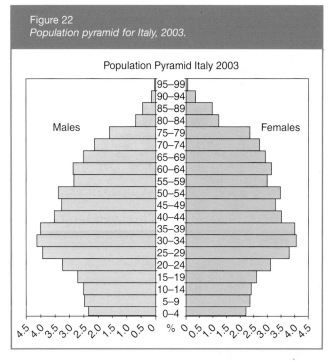

Figure 22
Population pyramid for Italy, 2003.

The population pyramid above outlines Italy's main problem – the smaller proportion of the population in the youth dependent population, and an increasingly elderly population. How is Italy's government going to cope with this increasingly

Figure 23
Population statistics for Italy.

	1970	1975	1980	1985	1990	1995	2000
Total population (thousands)	53,822	55,441	56,434	56,593	56,719	57,301	57,536
% 0–4	8.5	7.8	6.4	5.3	4.8	4.8	4.6
% 15–24	14.8	14.2	15.3	16.4	15.6	14	11.6
% 65+	10.9	12	13.1	12.7	15.3	16.6	18.1

	1970–75	1975–80	1980–85	1985–90	1990–95	1995–2000	
Pop. growth rate	0.59	0.36	0.06	0.05	0.2	0.08	
Crude birth rate	16.1	13	10.7	9.9	9.7	9.2	
Crude death rate	9.8	9.8	9.7	9.5	9.7	10.4	
TFR (child per woman)	2.33	1.89	1.53	1.35	1.28	1.21	
Life exp. (comb)	72.1	73.6	74.5	76.2	77.3	78.2	
Life exp. (male)	69.2	70.4	71.5	73.1	74	75	
Life exp. (female)	75.2	76.9	78	79.6	80.5	81.4	

elderly population, when there are not enough children being born to replace the decreasing working population?

Figure 23 outlines the main characteristics of Italy's population growth, which has slowed to an almost standstill from 1970 to the present day. At present the death rate is higher than the birth rate and the Total Fertility Rate (TFR) – the number of children per woman – is far below replacement level. This combined with an increasing life expectancy provides increasing problems for the Italian government. Italy is the first country in human history where the number of people over the age of 60 has surpassed the number of people under 20 years old. If the current trends remain the same, by the year 2035 the population over 60 will be greater than the number of the entire working population.

activities

1 What was the life expectancy in Italy in 1970 and 2000? [2]

2 By how much has the birth rate fallen in Italy between 1970 and 2000? [2]

3 Describe using information from Fig. 22 how the death rate has changed in Italy over time. [4]

4 Describe the changes to the aged dependent population in Italy over the past 30 years. [4]

HIGHER TIER

5 Considering the changes to the population structure of Italy outlined in questions 1–4, what are the consequences of such change on the following?

■ Services;

■ Employment;

■ Health care;

■ Education;

■ Pension provision;

■ Society. [12]

Population change in Mexico

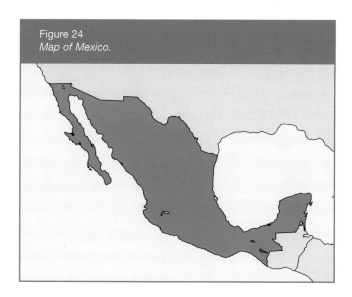

Figure 24
Map of Mexico.

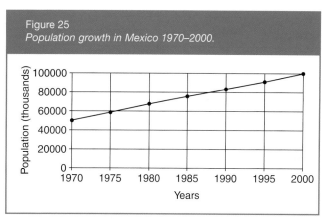

Figure 25
Population growth in Mexico 1970–2000.

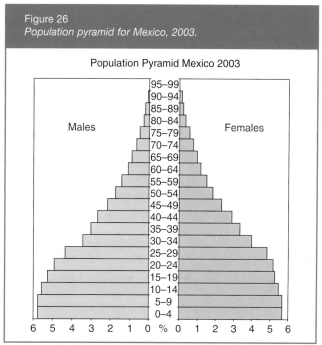

Figure 26
Population pyramid for Mexico, 2003.

Mexico is a country that has undergone massive population changes in the past 30 years. It is located just south of the USA but is very much regarded as its very powerful neighbour's poorer cousin. As can be seen from the graph below, the population of Mexico has doubled in the past 30 years.

The population pyramid above shows the extent of Mexico's problem – there is a large proportion of the population in the youth dependent population category. The government is going to have to find employment for this population as they move into the working population. Already it is of little wonder that Mexico City is the world's second largest urban

Figure 27
Population statistics for Mexico.

	1970	1975	1980	1985	1990	1995	2000
Total population	50,596	59,098	67,569	75,464	83,225	91,143	98,933
% 0–4	18.6	18.3	16.2	14.3	13.2	12.5	11.4
% 15–24	18.6	19.2	19.8	20.7	22	21.9	20.5
% 65+	4.3	4	3.8	3.7	4	4.3	4.8

	1970–75	1975–80	1980–85	1985–90	1990–95	1995–2000
Pop. growth rate	3.11	2.68	2.21	1.96	1.82	1.64
Crude birth rate	43.2	37.1	31.9	29.3	27	24.6
Crude death rate	9.5	7.7	6.4	5.7	5.2	5
TFR (child per woman)	6.52	5.3	4.24	3.61	3.12	2.75
Life exp. (comb)	62.4	65.1	67.5	69.6	71.2	72.5
Life exp. (male)	60.1	62.2	64.4	66.8	68.5	69.5
Life exp. (female)	65.2	68.6	71.2	73	74.5	75.5

area, with hundreds of thousands of people migrating to the city every year in the hope of employment. Unfortunately, the city and the government generally cannot cope with such rapid growth. As a result of so many people moving in, huge shanty towns encircle the city, providing substandard housing for millions of people.

Mexico's government has been working hard to reduce the growth rate of the population, and it can be seen from the table that this has been quite successful. The rate of growth has slowed down, and the number of children per woman has decreased significantly from 6.52 in 1970–75 to 2.75 in 1995–2000. However, because Mexico has such a large population of child-bearing age, the population is still growing at a rapid rate. This, and an increasing elderly population, means the government have a lot of people to support in the youth and aged dependent categories. Along with this, the fact that Mexico is a LEDC, with a low GNP, makes the situation even more difficult. Mexico faces potential problems in terms of its capacity to create sufficient new jobs to meet its national labour force. Many experts say that the next generation's decisions regarding where they will live and how many children they will have will make, or break, Mexico's future.

INTERNATIONAL MIGRATION

International migration is when people move from one country to another on a permanent basis.

As outlined previously, people who migrate are called **migrants**. Those who migrate **into** a country are called **immigrants** and those who **exit** a country are called **emigrants**.

There are many different reasons why people move from one country to another. There are decisions that people have to make based on **push** and **pull** factors. **Push** factors are **negative** reasons that would make someone want to leave a region. **Pull** factors are those that make someone want to move to a country. Pull factors can be considered the **opportunities** that the area has to offer.

While migration is defined as being a permanent move, there are different categories of migration. Some migration can be forced, whereas other migration is voluntary, some is permanent whereas other migration is temporary. Internal migration involves moving from one part of a country to another, within the country boundary.

There has been a lot of discussion in the news in recent times about migrants, with different terms being used to describe the movement of people into either the United Kingdom or the Republic of Ireland. Some of the most recent migrants to the

activities

1 What was the life expectancy in Mexico in 1970 and 2000? [2]

2 By how much has the birth rate fallen in Mexico between 1970 and 2000? [2]

3 Describe using information from Fig. 26 how the death rate has changed in Mexico over time. [4]

4 Describe the changes to the population aged 0–4 in Mexico over the past 30 years. [4]

HIGHER TIER

5 Considering the changes to the population structure of Mexico outlined in questions 1–4, what are the consequences of such change on the following?

■ Services;

■ Employment;

■ Health care;

■ Education;

■ Pension provision;

■ Society. [12]

Figure 28
Push and pull factors affecting migration.

War	Famine	Infertile soils	Mechanisation of farming	
University	Friends/family	High taxes	Medical provision	Peace
Terrorism	Religious persecution		Employment opportunities	
Climate	Retirement		Natural disasters	

activities

1 Sort the factors outlined in Fig. 28 into a table with two columns, one for push factors and one for pull factors.   [15]

2 Add as many push and pull factors of your own as you can.   [4]

3 Research the following terms which relate to migration and give a definition of each:

   (a) Permanent;

   (b) Temporary;

   (c) International;

   (d) Internal;

   (e) Voluntary;

   (f) Forced.   [12]

British Isles are **refugees** or **asylum seekers**. These people have left their homes because they are afraid of death or persecution in their own country. Figure 29 gives a definition of a refugee.

In 2001 there were 88,300 asylum applications submitted in the UK. These asylum seekers came from Afghanistan, Iraq, Somalia, Sri Lanka and Yugoslavia. In the same period, 10,330 asylum seekers lodged applications in the Republic of Ireland, with applicants coming from Nigeria, Romania, Moldova, Ukraine and the Russian Federation.

Economic migrants are people who migrate from one country to another to seek a better life. The host country is the country that accepts the migrants; the donor country is the area the migrants come from. Sometimes people migrate having applied to a country for a work permit or to a company for a specific job. This can have a very positive impact on the economy of the host and donor country. The host country may get either workers with skills that they require, or they may get a labour supply that is willing to do jobs that the other workers in the country are unwilling to do. The donor country obtains revenue from the workers who usually send money home to their families. Unfortunately for the donor country, however, they usually lose their most skilled and capable workers.

There are many different procedures that governments adopt to allow people to come into their country. The USA operates a system of work permits that allows people to work on a temporary or permanent basis. Permanent economic migrants to the US have to apply for a Green Card that allows them to work in the country. The Republic of Ireland has up until quite recently had an 'open door' system which allowed economic migrants to move to the country quite freely. This system is under review as there are fewer jobs available. One of the major problems a country faces when it invites migrants to the country during prosperous times is what to do when there are not enough jobs for the local population during less prosperous periods. Significant racial tension can arise if the local population believe that migrants have taken jobs that should be theirs. This happened in Germany following reunification in the 1990s. Former East Germans felt jobs that Turkish migrants had should have been theirs. This led to significant racial tensions.

Figure 29
Definition of refugee.

A refugee is a person who 'owing to a well-founded fear of being persecuted for reasons of race, religion, nationality, membership of a particular social group, or political opinion, is outside the country of his nationality, and is unable to or, owing to such fear, is unwilling to avail himself of the protection of that country ...'

The 1951 Convention relating to the Status of Refugees
www.unhcr.org

Sometimes people migrate to a country illegally. This can be very risky both for the migrant and an employer who knowingly employs an illegal worker. The migrant, if caught, can face imprisonment and/or deportation. The employer, although generally having benefited from employing someone for low wages, could also face legal consequences including a hefty fine.

Regardless of the means or decisions by which a person arrives in a country, the outcome of many people from different parts of the world moving to a region can be the creation of a **multicultural society**. This is a society that contains people from different cultural, racial, religious, linguistic or national backgrounds. Living in a multicultural society can bring about many **opportunities** and many **challenges**.

Some of the challenges or problems of living in a multicultural society can be experienced by both the immigrant community and the host community. These can include racial discrimination, physical and verbal abuse, employment problems, language difficulties, human right abuses, misunderstanding, fear, intolerance, lack of trust, suspicion and hatred.

Of course, there are also many features or positive aspects of living in a multicultural society which can be experienced by the migrants and the host community. These features include experiencing different foods, different restaurants, music, festivals, different types of entertainment and cultural events, as well as the opportunity to meet people from a range of different backgrounds.

Figures 30, 31 and 32 outline some of the positive and negative impacts of migration in the UK and the Republic of Ireland.

Figure 30
Migration Policy Review.

The Irish Minister for Justice has said a review of migration policy is needed to take account of the changing economic situation.

Michael McDowell said that Ireland had the most open immigration system in Europe in recent years and that it might now be appropriate to be more selective in terms of the skills and personal characteristics of prospective migrants, to ensure they matched the needs of our economy and our society.

The Minister said consideration would be given to the introduction of a green card system to allow a more structured approach to the selection of migrants.

Adapted from www.rte.ie/news

Figure 31
Policing in multicultural Ireland.

International police officers address a conference on multi-ethnic society

An international conference organised by the Gardaí opens in Delgany in County Wicklow this morning under the title 'Providing a police service in a Developing Multi-Ethnic and Multicultural Ireland'. The conference will be addressed by police officers from other European countries and from the US about their experiences of dealing with ethnic minorities. The conference takes place in the shadow of two surveys that showed the great majority of asylum seekers, especially black asylum seekers, had experienced abuse and racial discrimination. Racist leaflets and stickers have been sent to homes in Dublin's inner city, and the Lord Mayor has been subjected to abuse for extending the hand of welcome to refugees and asylum seekers.

Adapted from www.rte.ie/news/
Morning Ireland, Tuesday 27 April 1999.

Figure 32
Notting Hill Carnival.

The sights and sounds of the Caribbean filled the narrow streets of West London as thousands descended on the area to party at the 38th Notting Hill Carnival.

While up to 10,000 police officers prepared for 2 days of hard work keeping the event safe, visitors were enjoying the break from work to listen to steel bands, raga, soca, hip hop and UK garage.

Carnival goer Clark Ainsworth said: 'It's an amazing atmosphere. The streets are lined almost three deep now and everyone is enjoying the floats go by. The emphasis this year is definitely on culture, there are loads of steel bands and other types of music'.

Adapted from www.bbc.co.uk/news

activities

1. International migration has positive and negative impacts for both the host and donor areas. Complete the table below using the information in this section to outline the main consequences of migration. [8]

Consequences of migration	Positive	Negative
Host region		
Donor region		

2. Think about your own environment. To what extent would you consider that you live in a multicultural environment? Outline some of the more multicultural aspects of your area or your lifestyle. (You can consider here the influences of food or music in your daily life, even if your own environment is not really multicultural.) [6]

3. Complete two spider diagrams, one for challenges and one for opportunities that multicultural societies bring. Use the information outlined above as well as examples from the television, radio, Internet or newspapers. [8]

UNIT THREE
Population Growth and Sustainability

OVERPOPULATION

As stated previously **overpopulation** is when there are too many people in an area for the resources available to support an adequate standard of life. The term **resources** can be applied to water, farmland, fuel supply, housing or transport systems. In LEDCs an area is regarded as being overpopulated when people's living conditions fall below the most basic requirements. This can be seen where people live in shanty towns, face drought or starvation, or where the resources are **depleted**, i.e. their land is overcultivated or overgrazed and soil erosion takes place. Favelas (shanty towns) in Rio de Janeiro indicate that over 1 million of this city's inhabitants live in overpopulated conditions, the same could be said of the inhabitants of Kiberia, a shanty town in Nairobi in Kenya. Residents of these areas do not have adequate water supplies, sewage disposal, electricity, roads or appropriate housing standards. In both these situations, the circumstances could be improved if the governments were in a position to invest enough money to improve the living conditions of the people living in these areas. If this were done to a sufficient level, the area may no longer be considered overpopulated.

It can sometimes be difficult to decide if an area is overpopulated. In cities in MEDCs there may be a part of city life that could indicate overpopulation but other aspects that would not. London is a congested city with over 7 million inhabitants. Its inhabitants suffer traffic congestion, very high property prices and pollution, but generally there are enough resources for its inhabitants. In Tokyo, the largest urban area in the world, the Japanese manage to deal with very crowded conditions in imaginative ways, for example 'pushers' on the Tokyo underground systems squeeze as many people as possible into each train carriage!

There are some areas of the world that are **underpopulated**. This is when there are too few people in an area to make full use of the area's resources. This could mean that the governments of these countries would encourage people to migrate to the country to take up job opportunities. This happened in Australia in the 1960s and the 1970s, when large numbers of people migrated there from the British Isles. While underpopulation is not as bad a problem as overpopulation, the governments of underpopulated countries like Canada have to ensure that there will be enough in the working population in the future to provide taxes so that all the country's inhabitants could be provided for.

Since overpopulation is related to the number of people in an area and the resources that are available, if there is an imbalance, with either too few resources or too many people then governments have to adopt strategies that will try to even out any imbalance.

Figure 33 below outlines some approaches that countries have adopted to try and even out imbalances in different countries. Some may seem extreme and may lead to a consideration of the rights and wrongs of the approach adopted. However, it is important to remember that each approach was adopted in a particular context, within a culture that is often very different from that of Northern Ireland or Great Britain.

Figure 33
Population and resources.

Approach	Description	Method	Resource/population
China – One Child Policy	An attempt to slow the growth of China's population by allowing each family to have only one child.	Top down approach. Imposed by the government. The stated aim was to reduce the population for the benefit of the entire population. The policy was often enforced through coercion, financial penalties for individuals not conforming, with possible implications for the wider community.	Attempting to tackle the population issue. However, the effectiveness of the policy was reduced when Chinese citizens were able to improve their own standard of living through trade and industry, and were able to afford more children. In other words, they increased the resources available to them.
Mauritius – family planning policy	This island had a rapidly growing population so the government adopted a policy of encouraging couples to have fewer children, with the aim of reducing family size from the average six to two.	Through education and advice. Setting up family advice clinics and getting the support of the local community. Also, efforts were made to improve the resource base of the island, from being dependent on sugar cane, to tourism, and clothing manufacture.	Dealing with reducing the population as well as improving the resources available in the country. Overall, the policy has been successful, but has changed the social structure of the country significantly.
India – population policies since 1952	Financial incentives offered to people willing to undergo sterilisation with a peak of 8.3 million operations in 1977.	Indian government with a population rising towards, and now exceeding, 1 billion, adopted a policy of offering payments to people willing to undergo sterilisation. In 1993 96% of these operations were performed on women.	The government did offer other programmes including education and birth control clinics. India, by adopting a population policy in 1952, was the first LEDC to do so. However, the emphasis on sterilisation undermined peoples' confidence in the programme.
Burkina Faso – stone lines	An attempt made to stop increasing desertification through a small-scale/local approach using appropriate technology.	In areas vulnerable to soil erosion stones are laid down in a pattern that is roughly equivalent to contour lines. These stones trap moisture and prevent erosion.	This approach allows people to use appropriate technology to halt and reverse the process of desertification. This means the area is maintained and can be used to graze animals, increasing the carrying capacity of the land, therefore increasing available resources.

1 State the meaning of the term overpopulation? [2]

2 Why is it important to consider population and resources when discussing overpopulation? [3]

3 When is an area considered to be underpopulated? [2]

4 MEDCs often have different problems associated with overpopulation than LEDCs. What problems could overpopulation bring to a LEDC and what problems could overpopulation bring to a MEDC? [6]

HIGHER TIER

5 Study Fig. 33. State fully which of the strategies outlined you think is most effective in overcoming overpopulation? [5]

6 State fully which strategy you think is least effective. [3]

Figure 34
World wealth distribution.

The richest fifth of the world's population receives 82.7% of the world's total income and the poorest fifth receives 1.4%.

The North, with a quarter of the world's people, consumes 70% of the world's energy, 75% of its metals, 85% of its wood and 60% of its food. By 1990, more than 1.3 people lacked access to safe drinking water, 880 million adults could not read or write, 770 million had insufficient food for an active working life, and more than a billion lacked even the most basic necessities.

Today as then, an estimated 13–18 million people, mostly children, die from hunger and poverty each year. That is about 40,000 per day or 1700 an hour.

Adapted from *The Guardian*, Wednesday 19 February 1997

Figure 35
The global village.

'If the world were a global village of 100 people over 70 of them would be unable to read, and only 1 would have a college education. Over 50 would be suffering from malnutrition, and over 30 would live in what we call substandard housing. If the world were a global village of 100 residents, 6 would be from the United States. Those 6 would have half of the village's entire income; the other 94 would exist on the other half.'

Adapted from Peacemaking Day by Day Daily Readings
Pax Christi, Dublin.

CONTRASTS IN PATTERNS OF CONSUMPTION BETWEEN MEDCS AND LEDCS

As can be seen from Figs 34 and 35 there are differences between what the richest parts of the world have in terms of energy and resources and what the poorest parts of the world have. There is no doubt that MEDCs have the lion's share of the world's resources, and there are only limited efforts being made to try and distribute resources more evenly. As a result, the richer 'North' continues to consume much more in terms of the Earth's energy supplies and resources than the poorer 'South' does, even though there are more people in the LEDCs than in the MEDCs.

Figures 34 and 35 also demonstrate that the consequences for those living in LEDCs are lower standards of living, a poorer quality of life, unsatisfactory education and medical provisions, and even death from hunger and starvation.

The differences in consumption between MEDCs and LEDCs can be seen in Fig. 36, this shows social and economic data for the UK and Ethiopia. It can be seen that there are significant differences between

---

Figure 36
Indicators of consumption.

Indicators of consumption	UK	Ethiopia
CO_2 emissions per capita, 1998 (metric tons)	9.2	0
Commercial energy use per capita, 1999 (kg oil equivalent)	3,871	290
Forest area, change in 1990–2000 (1000 hectares)	170	−403
Number of vehicles per 1000 people, 2000	430	2
Urban population (%)	90	15

Figure 37
Explanation of variables.

Indicators of consumption	Explanation
CO_2 emissions per capita, 1998 (metric tons)	CO_2 is a gas that is emitted from car exhausts and power stations. The more CO_2 emissions there are the greater the amount of fossil fuels consumed.
Commercial energy use per capita, 1999 (kg oil equivalent)	Commercial energy use is the amount of energy that is used in trade and industry. The greater the figure the more industry there is and the greater wealth made for the country.
Forest area, change in 1990–2000 (1000 hectares)	Overgrazing, deforestation and soil erosion is a significant problem in the Sahel region of Africa, where many people rely on fuelwood for most of their energy needs. A minus figure indicates deforestation and a plus figure indicates afforestation – a more sustainable approach.
Number of vehicles per 1000 people, 2000	Car or vehicle ownership is an indicator of wealth. The wealthier the region, the more vehicles there will be. This will have implications for CO_2 emissions.
Urban population (%)	In urban areas, more resources need to be made available for the population. A lower urban population indicates a higher rural population. High rural populations usually indicate a high proportion of subsistence farmers.

the two countries in each of the variables. These differences exist because the UK is a MEDC with greater wealth, resources and services than Ethiopia, a LEDC.

The variables themselves show that there are differences. However, it is more important to understand why there are differences. Figure 37 illustrates what each variable indicates so that differences in the patterns of consumption between the two countries can be understood.

The UK, using large quantities of coal, oil, gas, petrol and diesel, has much higher levels of consumption than Ethiopia. The way of life in the UK, like other MEDCs, means that each person consumes massively more of the Earth's resources than people in LEDCs like Ethiopia.

activities

1 What percentage of the world's population lives in MEDCs? [1]

2 What percentage of the world's wealth does the middle 60% of the world's population receive? [1]

3 What percentage of the world's wealth does the richest 20% receive? [1]

4 How would you describe the distribution of the world's wealth? [3]

5 To what extent do you consider the world's resources to be evenly distributed? Use examples to support your answer. [5]

6 Keep a diary for 1 day and list all the resources that you use. These can range from water and food, to electricity for your hairdryer or stereo, as well as fuel to transport you from place to place. [8]

7 Now imagine you live in a country like Ethiopia. Complete a similar diary showing the resources you might consume in 1 day. [8]

8 What do the differences in the two diaries illustrate about the patterns of consumption between MEDCs and LEDCs. [4]

HIGHER TIER

9 For any three of the indicators outlined in Figs 36 and 37, give reasons for the differences they show in patterns of consumption between MEDCs and LEDCs. [6]

IMPACT OF ENERGY EXPLOITATION AND CONSUMPTION

Most of the energy that is consumed in MEDCs comes from fossil fuels. Fossil fuels are used whether it is petrol in our cars, coal, oil or gas in the power stations to generate our electricity, or diesel for the lorries, buses, cars and trucks that provide transport for goods and people all over the country.

Energy exploitation and consumption can have positive and negative impacts on people and the environment. Northern Ireland, unlike the rest of Great Britain, does not have a plentiful supply of natural energy resources. There is, however, one energy source it does have, and a debate is raging now in the 21st century regarding the proposed exploitation of Northern Ireland's own fossil fuel – lignite or brown coal.

Lignite in Northern Ireland

Lignite is a fossil fuel that could be considered a halfway stage between peat and coal. It is used primarily for electricity generation and there are many power stations in Poland and other eastern European states that use lignite. There are significant reserves of lignite found close to Ardboe on the western shores of Lough Neagh, as well as under the Lough itself, and there are also significant reserves found near Ballymoney in Co. Antrim. Since the 1980s there has been discussion whether or not to mine lignite in Northern Ireland. In the mid 1980s there was a very strong possibility that the government was going to allow mining to go ahead. At that time Meekatharra Minerals, an Australian mining company, wanted to build a power station and create an opencast mine near Ballymoney to extract the lignite and burn it in the neighbouring power station to generate electricity. In 2003 the debate opened up again. Meekatharra Minerals, now known as Auiron Energy, has submitted plans to mine the lignite and generate electricity. The hope is to mine 600 million tonnes of lignite to fuel a 600-megawatt power station to generate electricity. The lifespan of the mining operation and the power station would be 30 years. The impact to the local economy would be a £500 million investment in the project, the creation of 2000 construction jobs and 500 jobs in mining and in the power station.

Opencast mining involves the clearing of large areas of land. The 'overburden' – the soil and rock that covers the lignite – is taken away and the lignite is removed using huge diggers and trucks. Opencast mining does not involve the construction of tunnels and mineshafts like traditional mining operations, but it does mean that the surface of the mining area is significantly disturbed. It has been suggested that the mine in Ballymoney could be over 130 metres deep with a spoil tip (the removed overburden) over 1.5 km across. The entire operation is expected to extend 6 km by 5 km. One of the conditions of such mining operations is that the area is returned to its former state once the lignite is removed; however, the extent to which the area could ever be fully restored is questionable.

Obviously, such an operation has widespread implications for farmers and locals living near the proposed development, as well as for vegetation and wildlife.

There are also implications for electricity consumers who, faced with the possibility of obtaining cheaper electricity, would have a difficult decision to make whether or not to support the development.

WWF and Friends of the Earth have raised objections to the development. In addition, many local residents and businesses have objected to the plans. Some of their points of view are outlined below.

- There will be wholesale destruction of a mature and historic landscape which can never be recreated, and the loss of farms and family connections within the area.

- The proposed development would be incompatible with existing industry, i.e. agriculture and related industries.

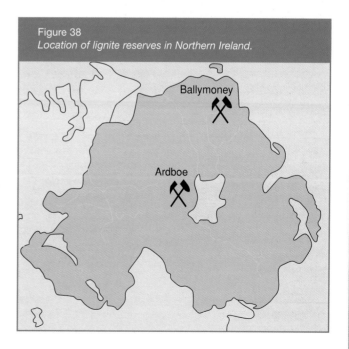

Figure 38
Location of lignite reserves in Northern Ireland.

Ballymoney

Ardboe

- There are concerns about the long-term health of those living in the wider area, especially the young and the elderly, due to the processes of opencast mining and emissions connected with power generation.

- Pollution: noise and dust, both atmospheric and visual, will result.

- An unacceptable increase in traffic on country roads with corresponding increases in road traffic accidents.

- Loss of local amenity and wildlife environment.

- The burning of lignite, a non-renewable source and one of the least efficient fossil fuels, will release significant amounts of CO_2 into the atmosphere thereby contributing to global climate change and at variance with the UK's carbon dioxide (CO_2) reduction policy.

Figure 39
Working at an opencast mine.

Figure 40
An opencast mine.

- Visual intrusion will affect the gateway to some of Northern Ireland's most famous tourist attractions.

While this type of energy exploitation and consumption has positive and negative impacts on people and the environment, the decision whether or not to go ahead with the development lies with the planners and the government. There is no doubt that no matter what decision is made, there will be many people dissatisfied with the outcome.

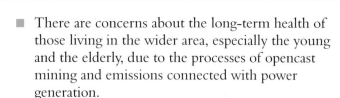

activities

1 What is lignite? [2]

2 Is lignite an renewable or a non-renewable resource? [1]

3 Why does Auiron want to exploit lignite at Ballymoney? [3]

4 State fully the possible positive impacts of the proposed lignite exploitation on

- The people of Ballymoney and Northern Ireland;

- The environment. [6]

5 What are the possible negative impacts of the proposed lignite exploitation on

- The people of Ballymoney and Northern Ireland;

- The environment. [6]

HIGHER TIER

6 Research the exploitation of other non-renewable resources that have positive and negative impacts on people and the environment. Think about the impact of oil exploitation and the problems illustrated by the sinking of the Prestige oil tanker off the coast of Spain in 2002. Another example could be the benefits the natural gas pipeline has brought to homes and industry in Northern Ireland. [6]

ONE RENEWABLE ENERGY PRODUCTION SCHEME AT REGIONAL/NATIONAL SCALE

Figure 41
Image of a wind turbine propeller.

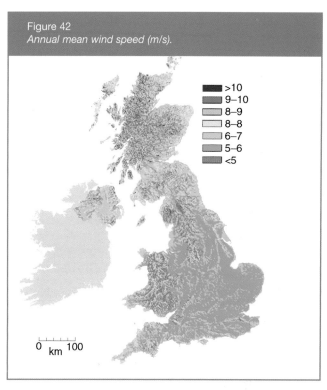

Figure 42
Annual mean wind speed (m/s).

▉	>10
▓	9–10
▒	8–9
░	8–8
▒	6–7
▒	5–6
▓	<5

0   km   100

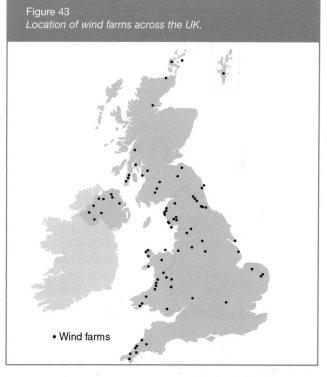

Figure 43
Location of wind farms across the UK.

• Wind farms

Wind power is being increasingly used as a means of electricity generation in western Europe. For example, the Republic of Ireland is currently developing an offshore wind farm of 200 windmills, which is being built on a sandbank in the Irish Sea south of Dublin. When this is completed, wind power will produce 10% of the country's energy needs and will be the world's largest offshore electricity-generating wind farm. One of the main benefits of this wind farm is that it will help allow Ireland achieve its target to cut greenhouse gas emissions by 13 million tonnes per year, as well as providing a more sustainable means of energy production.

Northern Ireland is also increasing the amount of electricity that it is generating by wind power. There are many regions in Northern Ireland that are ideal for the location of wind farms. (A wind farm is a group of wind turbines in the one location.) This can be seen from the map in Fig. 42, which shows the windiest parts of the British Isles. At present there are eight wind farms located in Northern Ireland and there are more being planned.

activities

1   What is the relationship between the location of the wind farms in the UK shown in Fig. 43 and the annual mean wind speed as shown in Fig. 42?   [4]

The company that develops wind farms across Northern Ireland is B9 Energy Services Ltd. It has been developing projects in Northern Ireland since 1992. The locations of some of its current wind farms are listed in the table below.

Site	Turbine capacity	Number of turbines
Rigged Hill Wind Farm, Co. Londonderry	500 kW	10
Corkey Wind Farm, Co. Antrim	500 kW	10
Elliots Hill Wind Farm, Co. Antrim	500 kW	10
Slievenahanaghan, Co. Antrim	1000 kW	1
Lendrums Bridge Wind Farm, Co. Tyrone	660 kW	8
Slieve Rushen, Co. Fermanagh	500 kW	10
Bessy Bell, Co. Tyrone	500 kW	10

The wind turbines on Northern Ireland's wind farms have towers ranging from 25 to 80 metres high. The rotors have a diameter of up to 65 metres. Figure 44 shows a typical wind turbine.

The turbines have three rotor blades that face into the wind and rotate around the horizontal hub at 28.5 revolutions per minute (rpm). The rotation of the blades drives a shaft in the nacelle (where all the electrical components are located), which is connected to a generator via the gearbox. It is the generator that makes the electricity.

Figure 44
A wind turbine.

Figure 45
A wind farm.

The towers of a wind turbine are made mostly of steel and are painted light grey because this is the colour most inconspicuous under most lighting conditions. The matt finish helps to reduce reflected light. Some towers are made of concrete. The blades are made of glass-fibre reinforced polyester or wood-epoxy.

One 660 kW wind turbine is able to produce enough electricity to meet the demands of approximately 350 households. Over the course of a year a wind turbine will produce about 30% of the amount of electricity it could theoretically have produced if it was working flat out all year long. On average, wind turbines produce electricity about 80–85% of the time depending on the location of the wind farm. If the turbines are not moving it is because there is too little or too much wind or they are being serviced. The turbine is designed to withstand gusts of up to 60 m/s.

As Fig. 45 shows, the best place to locate a wind turbine is on the top of smooth hills since wind speed increases with altitude, on open plains, along a shoreline and in mountain gaps.

Since wind power uses renewable energy it has many advantages over alternative energy supplies. For example:

■ Wind power produces no pollution.

■ Wind power off-sets the emissions of CO_2, NO_2 and SO_2 from the burning of fossil fuels.

■ Wind power is clean, free and infinite.

■ The price of wind power is not affected by fuel price increases or supply disruptions.

■ The land used for wind farms can be used for other purposes such as, farming, grazing and forestry.

However, while there are many advantages to producing electricity through the use of such a readily available and renewable source of energy,

Figure 46
Newspaper Report, January 29, 2003.

Wind-farm plan 'withdrawn'

A planning application for the controversial Carricknabrattoge wind farm near Lisnaskea, Co. Fermanagh has been withdrawn.

B9 Energy and Renewable Energy Systems withdrew their planning application for an 11-turbine wind farm in order to carry out further research after meetings with councillors and the Royal Society for the Protection of Birds (RSPB).

'Initial responses from the RSPB to the planning application made in October 2002 indicated that the proposed wind farm might affect the foraging behaviour of hen harriers in the Carricknabrattoge area', said B9 spokeswoman Denise Campbell.

She said the company was also aware of the concerns of the local community.

Campaigners, mainly residents from the Knocks community, were opposed to the siting of the wind farm on an area of blanket bog that they said was protected by EU directives.

Figure 47
News Report, 14 February, 2002.

Tourism fears over wind-farm plan

The north coast is famous for its sea views and some residents feel 'clean' energy would be generated to the detriment of tourism – one of the area's main incomes.

Don Wilmot who manages the Causeway Coast and Glens Regional Tourism Organisation has viewed offshore wind farms near Copenhagen.

He said he had heard reports that tourism had dropped by 40% in an area of Denmark with a lot of windmill development.

'Tourism is a major earner for the (north coast) region and generates some £100m of revenue', he said.

'Anything that would impact on us would give us serious cause for concern.'

Barbara Dempsey who owns an electrical shop on the sea front in Portstewart, said most people were 'adamant it would be a disaster' for the area.

'I, in line with a lot of local feeling, would be quite against something that would be detrimental to the view', she said. 'It may be ecologically sound, but certainly would do nothing for our tourist industry.'

But, her husband, Jimmy, would appear to be one of the few local people backing the wind farm idea. 'I would prefer for Northern Ireland or Ireland to have its own source of renewable power', he said.

---

there are many issues regarding its use. B9 energy has faced opposition to the construction of some of its wind farms in Northern Ireland. For example, Fig. 46 shows that wind farms, even though they are generally located in remote and upland areas, can cause conflict of interest between the developers and other interested groups.

B9 has also put forward a proposal to develop a wind farm off the north coast in the sea, between Portstewart in Co. Derry and the Inishowen Peninsula in Co. Donegal. This proposal could generate one-third of the energy output currently supplied by Kilroot Power Station. It would also help ensure that Northern Ireland reached its targets for energy production by renewable means. However, one of the main problems of the proposed development is the negative impact it would have on tourism in the area. Figure 47 outlines some of the points of view regarding the proposal.

While the production of electricity by renewable and sustainable means through wind power is generally considered to be environmentally friendly, it can be seen that there are also associated environmental problems. Different groups hold strong opinions about the benefits and problems associated with this issue, especially if they are directly affected by the development. Since wind farms have an environmental impact, those directly affected have to evaluate such schemes to determine whether they should go ahead or not.

Some websites useful to investigate for this topic

http://www.rte.ie/news/2002/1210/immigration.html

www.bbc.co.uk/news/ni

www.b9energy.co.uk

www.bbc.co.uk/education/revision/geography

www.georesources.co.uk

http://esa.un.org/unpp

www.unhcr.org

www.bbc.co.uk/news/race

www.bbc.co.uk/news/ni

www.prb.org

activities

1 Design a poster showing the benefits of wind power for Northern Ireland and the problems it brings. [8]

2 To what extent can electricity production by wind power be considered sustainable? [3]

3 Considering wind power as a means of electricity production across Northern Ireland, evaluate this method of electricity production in terms of the benefits and problems it brings to the environment. [8]

UNIT ONE
Economic Change Creates New Opportunities

ECONOMIC ACTIVITIES – DIFFERENT TYPES AND THEIR LOCATIONS

There are four types of economic activities:

■ PRIMARY

Primary economic activities involve people obtaining raw materials or extracting them from the environment (either the land or the sea); people who do this work have jobs in the primary sector of the economy.

■ SECONDARY

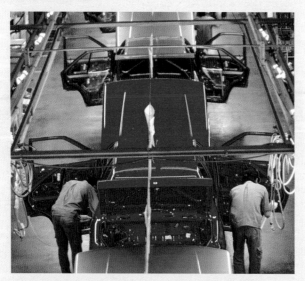

Secondary economic activities involve the processing of primary products to produce manufactured goods; raw materials from the environment are changed (e.g. by machines in a factory) into secondary products.

■ TERTIARY

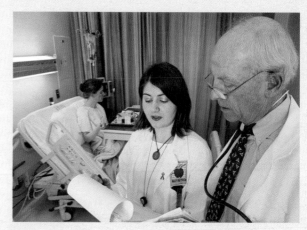

Tertiary economic activities provide a service to the public or help people in the community; some are essential services such as water and electricity supply, and these activities usually increase as a country becomes more wealthy.

■ QUATERNARY

Quaternary economic activities involve research and development into new designs and products, which usually require high levels of technology; they provide information and administrative services to other industries; often governments do not separate tertiary and quaternary activities in their statistics.

Economic activities are interlinked to make any product and get it sold to people. For example, to get a chocolate bar made from the processed raw materials, to get it packaged and distributed to customers, requires people who work in all four types of economic activity.

Figure 1
The outputs of the manufacture of a chocolate bar.

activities

1. Use Fig. 2 below to complete the following paragraph about the production of a chocolate bar. [4]

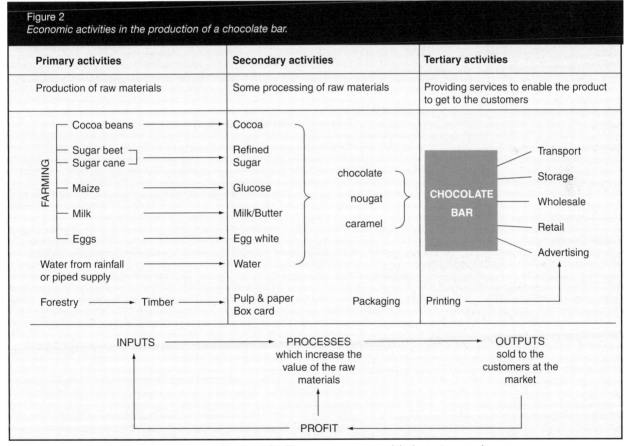

Figure 2
Economic activities in the production of a chocolate bar.

Primary activities	Secondary activities	Tertiary activities
Production of raw materials	Some processing of raw materials	Providing services to enable the product to get to the customers

FARMING
- Cocoa beans → Cocoa
- Sugar beet / Sugar cane → Refined Sugar
- Maize → Glucose
- Milk → Milk/Butter
- Eggs → Egg white
- Water from rainfall or piped supply → Water

chocolate / nougat / caramel

CHOCOLATE BAR → Transport, Storage, Wholesale, Retail, Advertising

Forestry → Timber → Pulp & paper Box card   Packaging   Printing

INPUTS → PROCESSES which increase the value of the raw materials → OUTPUTS sold to the customers at the market

PROFIT

The manufacture of a chocolate bar is like a system with inputs such as

_____. Some of these such as cocoa, sugar, milk and eggs are

_____ _____ which are combined together and processed to make the chocolate

bar. In order for the processed chocolate bars to reach their market, they rely on tertiary

economic activities such as _____.

continues

2 Sort the following activities into the four groups – primary, secondary, tertiary and quaternary.   [4]

tourist office   fishing
leisure centre   software design
quarrying   car assembly
microchip design   mining
product development   weaving linen
farming   refining oil
transport   scientific research
health care   spinning cotton

3 Add one more economic activity to each of the four groups.   [4]

4 Study the graph in Fig. 3 showing the changes in employment in different economic activities in the UK. Answer the following questions.

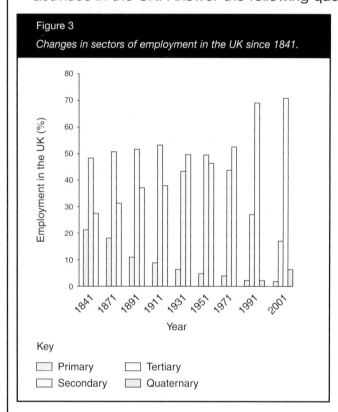

Figure 3

Changes in sectors of employment in the UK since 1841.

Employment in the UK (%)

Year

Key

☐ Primary   ☐ Tertiary
☐ Secondary   ☐ Quaternary

(a) State the year when:

  (i) employment in primary activities fell below 10%;

  (ii) secondary employment was at its highest;

  (iii) employment in tertiary activities became the greatest;

  (iv) quaternary employment was first indicated.   [4]

HIGHER TIER

(b) Suggest reasons for:

  (i) the decline in primary employment;

  (ii) the rise of tertiary employment.   [6]

ACTIVITIES USING ICT

Different countries have different proportions of their labour force employed in the three main economic activities as shown in the table below.

	France	UK	Argentina	Taiwan	Botswana	Kenya
% primary	6	2	6	4	28	81
% secondary	29	22	31	41	11	7
% tertiary	65	76	63	55	61	12

5 Construct a pie chart for each country (or one graph with parallel bars for all countries) using the table of percentages employed in different sectors.   [6]

6 Describe the differences between MEDCs and LEDCs.   [6]

7 Suggest reasons for these differences in types of employment.   [6]

INDUSTRIAL LOCATION

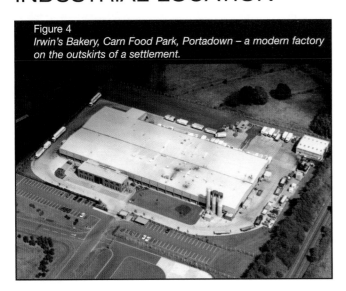

Figure 4
Irwin's Bakery, Carn Food Park, Portadown – a modern factory on the outskirts of a settlement.

Industrial location means the place chosen for an economic activity; these locations can vary from rural to urban. Some examples are:

- A mill grinding corn in a rural area;
- A carpet factory in an **inner-city area**;
- A food-processing company in a specially built industrial estate on the outskirts of a city.

Factors affecting the location of industry

If a manufacturing industry is going to make a profit, the owners must consider the **location** for their factory; they will choose a location for the factory that keeps their profits high and keeps costs down.

Physical factors

- Raw materials – the more heavy and bulky these are, the closer the factory must be located to the raw materials. Northern Ireland does not have many local raw materials (except agricultural produce) so most of the raw materials for industries have to be imported through docks at Belfast, Larne, Londonderry and Warrenpoint.

- An energy supply – power is needed to work machines in a factory. In Belfast, power was once supplied by fast-flowing rivers such as the River Farset which flowed from the Antrim Plateau through west Belfast and provided water power for linen mills; power then was supplied from burning coal, so Belfast became an important industrial centre because coal could be imported through the docks; nowadays electricity can be transmitted long distances so factories do not need to be located close to a source of power.

- A large site with flat land on which to build the factory – in the past this was available in the inner-city zones of settlements, but now large areas of flat land for factories are only available on the edge of towns and cities, often on especially built industrial estates.

Human factors

- Access to a large market where goods can be sold – this is usually the towns and cities where a large number of people are living, e.g. the main market for goods in Northern Ireland is the more densely populated area of Greater Belfast and other large towns. However, Northern Ireland has a small population, so many industries must export goods to mainland Great Britain or to Europe or elsewhere. Easy access is important in getting products to the market so good transport links are required.

- Capital or money must be available – in the past wealthy entrepreneurs invested in industry in Northern Ireland, e.g. the Mulholland brothers and businessmen from Scotland invested in the linen industry; now capital is borrowed from banks or is obtained from government grants.

- The availability of a workforce – this may need to be skilled if the processing work requires the use of machinery, a high level of technology and/or people who are able to design and improve products; workers with good standards of education are usually found in towns and cities with universities and colleges.

- Transport costs – these vary depending on whether the raw materials and/or the finished processed products are bulky or light, fragile or perishable. In the past, transport was by river and canal or railway, e.g. in the past goods were easily transported to Portadown along the Newry Canal; now road transport is more frequently used, so access to motorways and main roads is important, e.g. the A8 from Belfast to the port of Larne.

In the past many traditional industries needed large amounts of raw materials and power supplies, and so coalfields were common locations. In the UK, iron and steel and engineering industries were all located on the coalfields. Nowadays industries are more footloose; this means they can locate anywhere. This is because:

- Electricity is available in most places and is transmitted everywhere on the super grid;
- Transportation has improved so industries do not need to locate close to raw materials;

Figure 5
Factors affecting the location of industry.

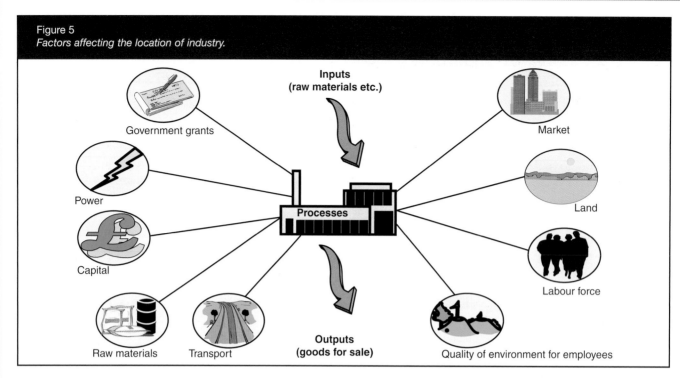

- The factories make products that are often light in weight but expensive to buy.

Sometimes there are several locations with roughly similar costs and so other factors become more important to the location of modern industry, such as:

- The availability of government grants at one location;

- The quality of life for the employees of the company; this includes factors such as attractive scenery and good educational and recreational facilities for the workers and their families.

Figure 6
The influence of three factors on the location of industry

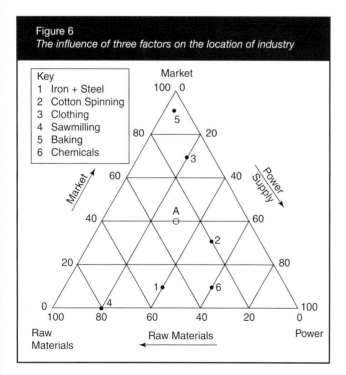

activities

1 (a) (i) Name one industry located close to its market.

(ii) Name one industry located close to a power resource.

(iii) Name one industry located close to the raw materials it needs. [3]

(b) Mark onto the triangular graph (shown in Fig. 6) the following industries using a suitable key:

printing newspapers

flour milling using imported grain

glass works using furnaces at high temperatures

brick works using much clay [4]

2 Explain the location of Industry A on the triangular graph. [3]

3 Suggest two other factors and explain how each influences the location of industry today. [6]

HIGHER TIER

4 Using the two maps in Fig. 7 explain how the **location of industry** in the UK has changed. [5]

continues

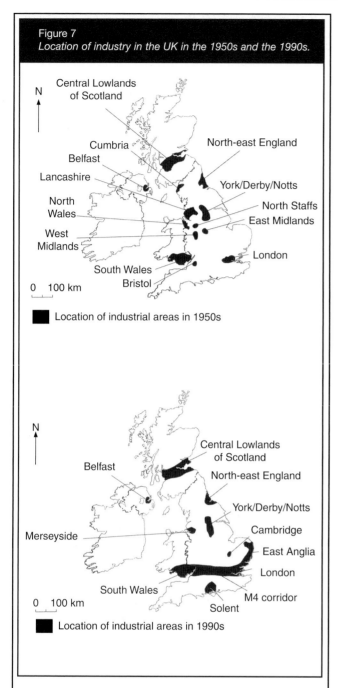

Figure 7
Location of industry in the UK in the 1950s and the 1990s.

Central Lowlands of Scotland
Cumbria
Belfast
Lancashire
North Wales
West Midlands
North-east England
York/Derby/Notts
North Staffs
East Midlands
South Wales
Bristol
London
0   100 km

■ Location of industrial areas in 1950s

Belfast
Central Lowlands of Scotland
North-east England
York/Derby/Notts
Cambridge
Merseyside
East Anglia
South Wales
London
M4 corridor
Solent
0   100 km

■ Location of industrial areas in 1990s

5   Describe how the location of Irwin's Bakery (shown in the Fig. 4) is attractive for industry. [6]

6   (a)   On an outline map of Northern Ireland mark the location of an industrial area. [2]

(b)   Annotate the map of Northern Ireland to show the main factors influencing location of industry in the area you have marked. [6]

continues

THINKING SKILLS ACTIVITY: ECONOMIC CHANGE

1.   farm
2.   linen mill
3.   cinema
4.   tyre factory
5.   people available to work
6.   coal mine
7.   oil rig
8.   competition from other countries
9.   customers
10.   shipyard
11.   machines replace people
12.   hospital
13.   quarry
14.   power for machines
15.   resources run out
16.   leisure centre
17.   car factory
18.   school
19.   competition from other industries
20.   raw materials

Task 1
Each of the numbers in the sets of three relates to the topic above. Can you work out with your partner which is the *Odd One Out* and what connects the other two?

Set A	1	7	10
Set B	17	4	14
Set C	3	2	12
Set D	20	7	13
Set E	11	15	18
Set F	9	5	15
Set G	3	5	16

Task 2
Still with your partner, can you find *one more* from the list to add to each of the sets above so that *four* items have things in common, but the *Odd One Out* remains the same? Think about why you have chosen each one.

Task 3
Now it's your turn to design some sets to try out on your partner! Choose three numbers that you think have something in common with each other and one that you think has nothing to do with the other two. Get your partner to find the *Odd One Out*, then do one of theirs. Try a few each, but remember to be reasonable.

Task 4
Can you organise all the words into groups? You are allowed to create between three and six groups, and each group must be given a descriptive heading that unites the words in the group. Try not to have any left over. Be prepared to rethink as you go along.

Source: *Thinking Skills*, N.I. Education Board (1999).

CHANGE IN FUNCTION OF INDUSTRIAL PREMISES

Industrial premises are the buildings used by economic activities. Many industries began in locations that are no longer suitable for a modern factory. Old inner-city locations may be congested, with streets too narrow for modern articulated lorries and there may be no space to expand the factory or provide parking for the workforce. The buildings may be too old to renovate to suit modern methods of production. Factories in these locations may close down or move out to locations outside the town or city on industrial estates. This gives opportunities to redevelop the site and change the use of the land at that location.

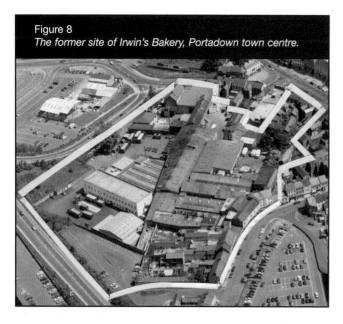

Figure 8
The former site of Irwin's Bakery, Portadown town centre.

Figure 9
The High Street Mall in Portadown.

LOCAL/SMALL-SCALE STUDY OF THE CHANGE IN FUNCTION OF INDUSTRIAL PREMISES – YORKGATE, BELFAST

Yorkgate was the site of the Gallagher's Tobacco factory until 1988. It is now a retail and leisure complex. This is an example of a change in function from secondary manufacturing industry to tertiary economic activity. It is also an example of **sustainable** economic **development** because an inner-city 'brownfield' site has been redeveloped and reused instead of building on a new 'greenfield' site on the edge of the city.

Gallagher's Tobacco factory was built at York Street and was a large cigarette-making company; it was a secondary manufacturing industry and once employed about 1000 people. It was closed in 1988. The reasons for the closure were:

- People have changed their smoking habits and are aware of the dangers of smoking and want to follow a more healthy lifestyle.

- The factory lost the market in which to sell their cigarettes – more of their customers were abroad instead of in the UK

- It became cheaper to manufacture cigarettes in factories in LEDCs where labour costs are lower.

Yorkgate is a retail and leisure complex with 31,500 square metres of space available; it is occupied by different functions. There are plans to expand this complex by adding to its retail space and providing further car parking facilities for customers.

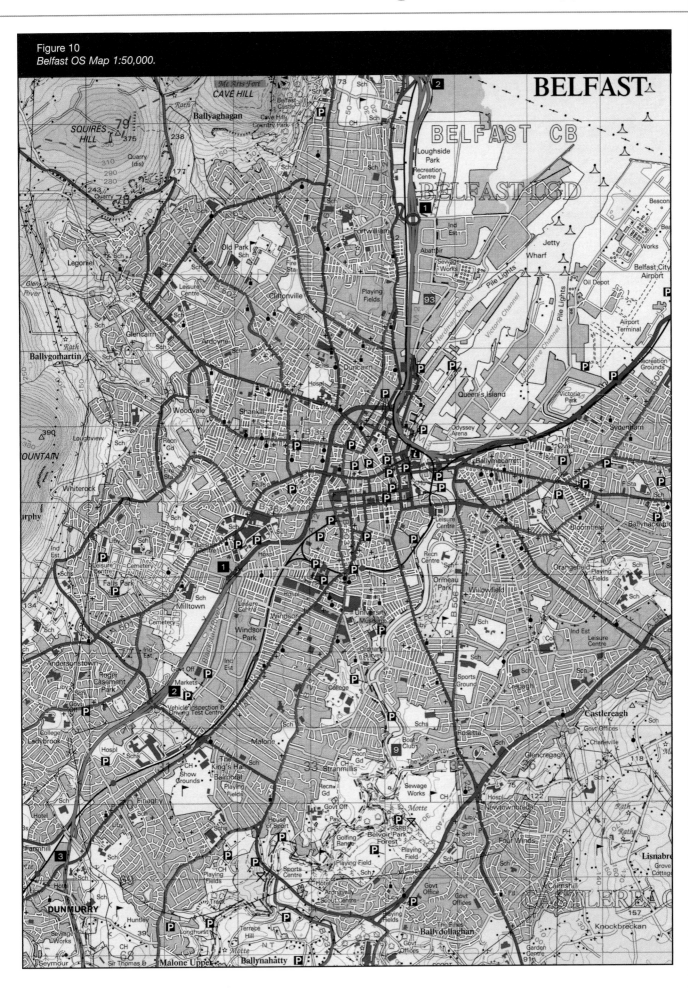

Figure 10
Belfast OS Map 1:50,000.

Map work activities on the location of Yorkgate (using Figs 10–13)

Copy and complete the following using the information on Yorkgate.

This site occupies 13 acres on three floors between the following streets:

York Street to the east, _____ _____ Street to the west, Brougham Street to the north and _____Street to the south.

Factors influencing the location of Gallagher's Tobacco factory:

- Gallagher's was located in the _____ _____ zone to the _____ direction of the CBD

- The factory was only a short distance from the CBD about _____ km.

- The factory had a pool of labour easily available; this came from …

- The distance from this site to the docks is only _____ km. This was convenient for a tobacco factory because …

Other factors have become more important to the location of the Yorkgate retail and leisure complex:

- Accessibility by road is excellent at this location because …

- The railway station is only _____ km away and this is good for the complex because …

- The complex is easily accessible to customers because …

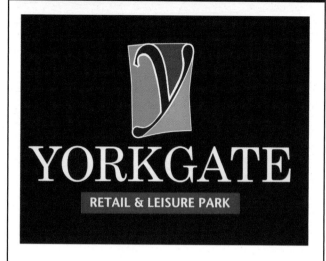

Figure 11
List of tenants at Yorkgate.

TESCO STORES	CARPENTER & MCALLISTER
TRAVELCARE	
EYECARE OPTICIANS	BURGER KING
BOOTS	BIG W
JAMES ALEXANDER (JEWELLERS)	PIZZA HUT
THE SPORTS COMPANY	LUNAR LAND/ TWILIGHT ZONE
SELECT	GALAXY BINGO
YORKGATE TAXIS	GALLAGHER'S BAR
BARRETTS FACTORY OUTLET	THE MOVIE HOUSE
HARVEYS	ARENA HEALTH AND FITNESS CLUB
CARPETRIGHT	HARRY RAMSDENS

Location factors for the present use of the site at Yorkgate:

- **Infrastructure** and accessibility: infrastructure, which means the network of services to benefit people, industries and businesses, e.g. telephone lines, water supply and sewage disposal. A major item of infrastructure is good transport links.

- Employees: Yorkgate employs about 350 people in service or tertiary jobs; many of these are part-time workers, students and casual workers, and many of them live near the complex.

- Customer catchment area: 550,000 people live within 15 minutes' drive time of Yorkgate.

- Space for parking: Yorkgate is on a large site and has a lot of space for the many customers who drive to the complex.

Figure 12
Location photograph and information on Yorkgate.

Situated within half a mile of Belfast City Centre

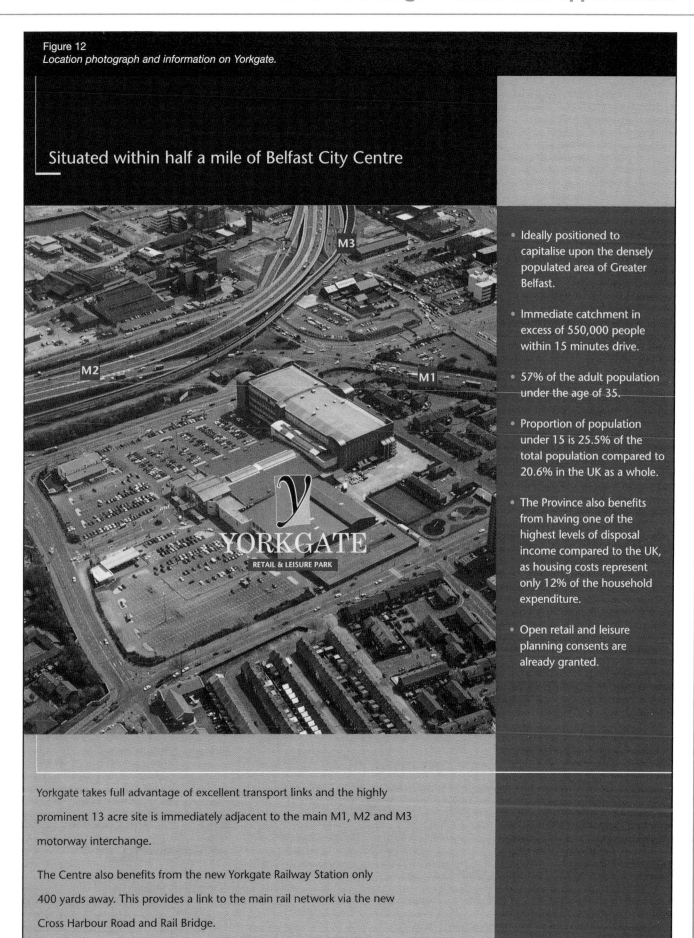

- Ideally positioned to capitalise upon the densely populated area of Greater Belfast.

- Immediate catchment in excess of 550,000 people within 15 minutes drive.

- 57% of the adult population under the age of 35.

- Proportion of population under 15 is 25.5% of the total population compared to 20.6% in the UK as a whole.

- The Province also benefits from having one of the highest levels of disposal income compared to the UK, as housing costs represent only 12% of the household expenditure.

- Open retail and leisure planning consents are already granted.

Yorkgate takes full advantage of excellent transport links and the highly prominent 13 acre site is immediately adjacent to the main M1, M2 and M3 motorway interchange.

The Centre also benefits from the new Yorkgate Railway Station only 400 yards away. This provides a link to the main rail network via the new Cross Harbour Road and Rail Bridge.

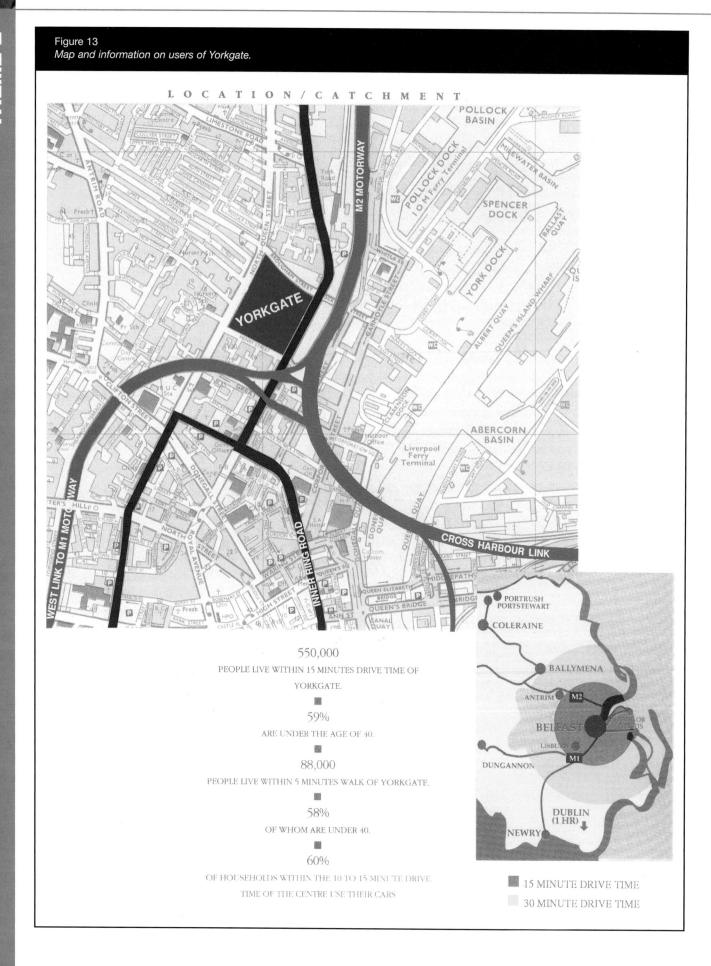

Figure 13
Map and information on users of Yorkgate.

LOCATION / CATCHMENT

550,000
PEOPLE LIVE WITHIN 15 MINUTES DRIVE TIME OF
YORKGATE.

59%
ARE UNDER THE AGE OF 40.

88,000
PEOPLE LIVE WITHIN 5 MINUTES WALK OF YORKGATE.

58%
OF WHOM ARE UNDER 40.

60%
OF HOUSEHOLDS WITHIN THE 10 TO 15 MINUTE DRIVE
TIME OF THE CENTRE USE THEIR CARS

15 MINUTE DRIVE TIME
30 MINUTE DRIVE TIME

activities

Answer these questions using Figs 10–13.

1. From the list of tenants at Yorkgate find examples of the following functions:

   Retail, Leisure, Restaurants, and other services. [4]

2 Describe how Yorkgate has a good transport infrastructure. [3]

3 Explain why this complex is in a good location to attract people to use it, using the resources provided. [6]

4 Suggest two advantages of such a complex to the local people and to the economy of the area. [6]

Advantages of Yorkgate to people

- The derelict building that was the cigarette factory has been given a new lease of life as it is now part of the complex and is no longer an eye-sore to the local inhabitants in the streets around the complex.

- There are services and amenities such as shopping outlets, leisure facilities, restaurants and a cinema provided within the complex; these serve the local people who live in the nearby houses; they have easy access to their work and to the shopping and leisure facilities, etc.

- There is a variety of flexible types of work available in the complex. People can work shifts, evenings or part-time; this suits different types of people such as young adults seeking jobs after school or mothers with young children. Many of the workers and customers live within walking distance of the complex (see Fig. 13 for a map of the area for names of streets and housing areas).

- Many of the people who live within walking distance are young; having the complex provides alternative ways to occupy their leisure time; it gives them opportunities to get part-time jobs or to use the services such as the cinema.

Advantages of Yorkgate to the economy

- There are jobs available for local people who live in this inner-city zone; most of the workers live in housing areas within walking distance such as York Road, Duncairn and New Lodge Road.

- Service industry located on this site earns greater income from the land than its value as a secondary manufacturing industry.

- There is a large threshold population of 550,000 people living within 15 minutes' driving time of the complex; this means there is a large threshold population to support the functions on the site and to make a large profit.

- The people living in Northern Ireland have lower housing costs than other places in the UK, so this means they have more disposable income and so can afford to use the complex often and this earns it more income.

FACTORS OF LOCATION OF HI-TECH INDUSTRY

During the 20th century many of the UK's traditional industries went into decline due to a depletion of power resources, production of outdated products, location in poor sites such as congested inner-city areas, lack of **investment** and foreign competition. However, economic changes and advances in **technology** have led to more automated systems in, for example, industry and office environments, and more widespread use of computers and electronic devices at home and in the workplace, and this caused the hi-tech industry to grow very rapidly in the 1980s and early 1990s.

Hi-tech industries are those industries that make sophisticated products incorporating micro-electronics and require a great deal of scientific research and development. These industries spend at least 5% of their annual turnover on research, development and design.

Examples of hi-tech industry

Examples of hi-tech industries are those associated with consumer electronics and those relating to electronic equipment.

Figure 14
Hi-tech products.

Consumer electronics	Electronic equipment
Electronic games	Computers
Televisions	Telecommunication devices
Video cassette (VCRs)	Control systems for industry
Radio receivers	Office equipment
DVD players	Equipment for testing and measuring
Hi-fi equipment	Equipment for biotechnology/pharmaceutical industry
Pocket calculators	Equipment for aerospace and the military
Tape recorders	Microelectronics for incorporation in consumer products, for example washing machines, cars and burglar alarms.

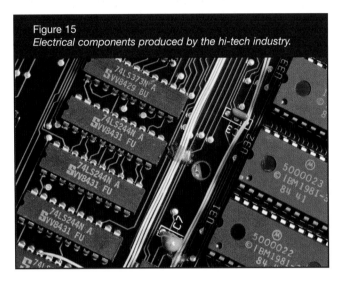

Figure 15
Electrical components produced by the hi-tech industry.

Hi-tech industries can be subdivided into:

- The *sunrise industries* which have a high-technology base.

- *Information technology industries* which involve computers, telecommunications and micro-electronics.

Hi-tech industries are often described as *footloose industries* in that their location is not affected by the traditional location factors such as being close to raw materials or power sources. However, hi-tech industries cannot locate simply anywhere as they have specific locational requirements, and this is why they often cluster together. Once an area has become attractive to a firm it tends to attract more companies of a similar nature leading to agglomerations or concentrations of industries. By locating close to each

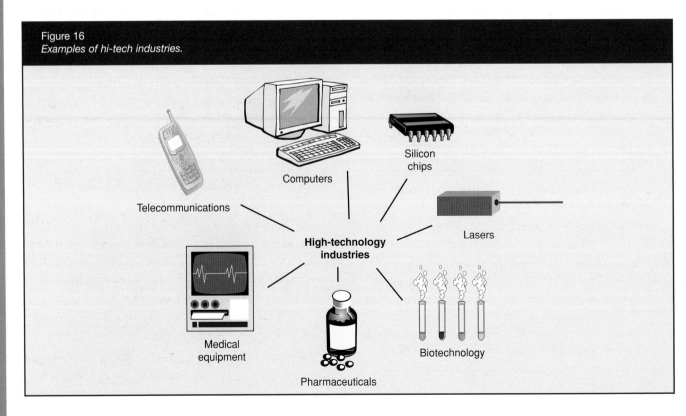

Figure 16
Examples of hi-tech industries.

Telecommunications

Computers

Silicon chips

Lasers

High-technology industries

Medical equipment

Pharmaceuticals

Biotechnology

other hi-tech firms can exchange ideas and information, and share amenities such as the communications **infrastructure**. They also have the advantage of being able to share maintenance and support services. By providing services and support for each other the industries help each other to grow and produce a multiplier effect. This means that the initial investment in economic activity in a region has beneficial knock-on effects elsewhere in the economy.

There are two main types of hi-tech industry:

- **Quaternary hi-tech industries** that develop products through research and development. These industries require skilled scientists, engineers and technicians, and so they tend to locate close to a highly skilled labour force and near a research centre such as a university. Science parks, which are industrial estates with links to universities, provide popular locations. They have attractive layouts with grassy areas, ornamental gardens and ponds. Hi-tech firms developing new products are often therefore located in these environmentally attractive areas, for example the out-of-town greenfield sites of university towns. Such locations may be considered attractive from a climatic, scenic, health and social point of view, and so provide a pleasant environment in which to work. In the UK the largest cluster of hi-tech industries involved in product development is in Silicon Fen near Cambridge.

- **Secondary hi-tech industries** that specialise in the manufacture of products by the assembly of components and raw materials. These industries are often found close to a cheap labour force, which may be skilled in the precision assembly of small parts. In MEDCs such as the UK they locate in peripheral regions where government financial incentives are available. The largest concentration of hi-tech production industries in the UK is in Silicon Glen in central Scotland.

Also associated with hi-tech industry are those companies which market the products and deal with the administration of the company. These industries prefer a city location or proximity to a city where there is a workforce with good managerial skills. Many of these industries are found in south-east England close to London.

The location of hi-tech industries in the British Isles

Concentrations of hi-tech electronics industries in the British Isles include:

- The Central Lowlands of Scotland;
- Around Cambridge;
- Along the M4 Corridor and into South Wales;
- Around Antrim in Northern Ireland;
- Numerous sites in the Republic of Ireland, for example the Science Parks at Galway linked to University College Galway and the Co-operative Education Centre and Science Park at Limerick.

The Central Lowlands of Scotland – Silicon Glen

Scotland's Central Valley has been called Silicon Glen because of the number of electronics factories located there. Many overseas companies such as Seiko, IBM and Texas Instruments can be found in this area. These factories are primarily concerned with the assembly of raw materials. Factors which have encouraged industries to locate in this area are:

- Cheap labour available due to the decline of traditional heavy industry such as iron and steel and shipbuilding;
- The local labour force is skilled in precision engineering due to experience in the declining heavy engineering industries;
- There are good transport links, for example Glasgow airport, rail connections, and road links via the M8, M9 and M90;
- Initially there were attractive government incentives and grants to industries setting up in this region;
- Proximity to attractive highland scenery.

Around Cambridge – Silicon Fen

This area has many companies involved in research and development located, for example, in Cambridge Science Park and St John's Innovation Park, Cambridge. The area attracts industry due to:

- Excellent research links with Cambridge and other universities;
- Proximity to M11 corridor and alongside the A10;
- Good access by rail and air, for example Stansted airport is 48 km from Cambridge. It is also within good travelling time by road to Heathrow (1.5 hours) and Gatwick (2.5 hours);
- Land values are lower on edge of urban area and there is room for expansion.

The project to develop Cambridge Science Park began in 1970 when the owner, Trinity College, Cambridge,

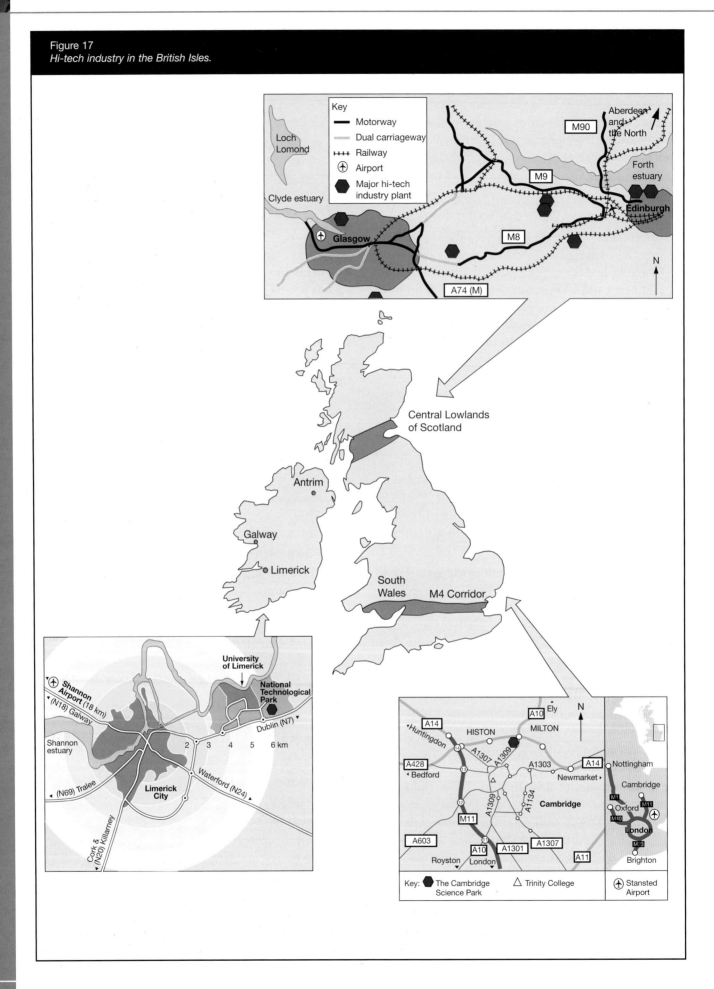

Figure 17
Hi-tech industry in the British Isles.

Figure 18
Cambridge Science Park.

decided to use its own lands and funds to provide a pleasant location for local hi-tech industries and research institutes. The park has flourished and by 2000 the cluster of hi-tech industries in the area had grown to 120 companies employing around 35,000 people. The demand for space has increased dramatically and a further 8.8 hectares of land is currently under development adjacent to the park. The park has many excellent amenities such as a state-of-the-art conference centre, restaurants, the Q.ton forum for conferencing and leisure, the Revolution fitness centre and a 115 child-place nursery. Life sciences such as pharmaceuticals are the dominant sector along with internet and telecommunication-related companies. Companies located here include Epson (UK), Toshiba Research, Sun Microsystems and Ionix Pharmaceuticals.

The Antrim Technology Park

The Antrim Technology Park was set up in 1986 by the Industrial Development Board for Northern Ireland (IDB) to meet the specific requirements of hi-tech based industries involved in the production of computer hardware, software development, electronics, light engineering, healthcare, research and development, and design. Accommodation is provided in various sizes up to 1750 m^2. There is also a large conference suite with audio visual equipment and catering facilities. Hi-tech companies located at the park include Dart (NI), Amtec, Parity Solutions

Software factory (Ireland), Fujitsu Communications (Ireland) and Schraderer Electronics Ltd.

The location was chosen and industries have been attracted to the park for the following reasons:

- Existence of a well-educated and highly motivated workforce;

- Availability of government or EU financial assistance, e.g. AFA Dart received £160,000;

- The park is in close proximity to the research facilities and industrial support services of the Queen's University at Belfast and the University of Ulster at Jordanstown;

- An attractive location of 32 hectares in a woodland setting;

- Excellent communications, for example Belfast International Airport is 5 km away and the port of Larne and city of Belfast are easily accessed by motorway (M2).

The Plassey (National) Technological Park, Castletroy, Limerick

The 263 hectares National Technological Park located beside the University of Limerick has a thriving cluster of more than 80 hi-tech firms which employ 5400 people. The park has an innovation centre which provides companies with access to broadband connectivity and managerial support. Dell Computers, Analog Devices, NetG and

Clarus (Telecommunications), among others have made Limerick their European Headquarters.

Factors which attracted hi-tech firms to this area include:

- Shannon International airport is located 20 km from Limerick city. Limerick/Shannon has been designated a Gateway in the National Development Plan and proximity to Shannon airport increases the development opportunities;

- National road improvements mean good access. As there is a well-developed infrastructure, the park is within easy reach of Cork, Galway and Dublin by road or rail;

- The University of Limerick is renowned internationally for producing graduates in the areas of engineering, science and information technology, and also for its research and development capabilities. This provides a pool of highly motivated, educated, skilled graduates;

- There is also good labour availability in the area as Limerick is Ireland's third city with a population of 52,000 and another 25,000 people live in the suburbs;

- State incentives towards business development. There is substantial government investment in Research, Technological Development and Innovation Initiatives;

- The area has broadband connectivity and this has been given priority in the city's development action plan, to be developed further giving increased capacity;

- Spare capacity in the park offers the opportunity for expansion.

activities

1 Draw a simple outline map of an area of hi-tech industry in Northern Ireland, for example the Antrim Technology Park. Annotate your map to show the main factors attracting hi-tech industry to this location. [10]

ICT

2 Research the locational factors attracting hi-tech industry to the M4 corridor from London to South Wales (Sunrise Strip). Find information on:

- The main communications links – roads, railways and airports;

- University and research centres in the area;

- The location and names of some of the industries;

- Nearby areas of attractive scenery;

- Availability of labour. [12]

web link & extra resources

www.cambridge-science-park.com/home.html

www.antrim.gov.uk

www.bionorthernireland.com

www.biotech-inst.com

www.ulst.ac.uk

UNIT TWO
The Impact of Global Economic Change

GLOBALISATION

Many goods can be bought and sold in a huge number of countries in the world. Examples of brand names which are famous worldwide are Coca-Cola, Levi Jeans, McDonalds and Nike trainers. Think of the origin of many items in your home, e.g. food, clothes, music, TV soaps, movies, etc. These are all examples of how **globalisation** has affected people around the world. People, goods, money and ideas move around the world faster and more cheaply than ever before. Most of the market in major products is controlled by only a few companies as shown in Fig. 19.

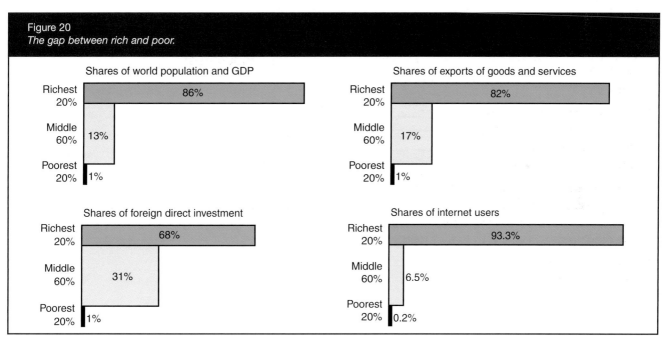

Figure 19
The market share of the top ten corporations by sector.

Telecommunications 86%

Pesticides 85%

Computers 70%

Veterinary medicine 60%

Pharmaceuticals 35%

Commercial seed 32%

Figure 20
The gap between rich and poor.

Shares of world population and GDP

Richest 20% — 86%
Middle 60% — 13%
Poorest 20% — 1%

Shares of exports of goods and services

Richest 20% — 82%
Middle 60% — 17%
Poorest 20% — 1%

Shares of foreign direct investment

Richest 20% — 68%
Middle 60% — 31%
Poorest 20% — 1%

Shares of internet users

Richest 20% — 93.3%
Middle 60% — 6.5%
Poorest 20% — 0.2%

Globalisation means:

- That companies operate in many countries and so international operations are increasingly important for people and companies;

- Decisions taken in one country can quickly and easily affect other countries because of the improvements in transportation and the spread of global communications, such as e-mail, fax and video-conferencing;

- Economic power is concentrated in the hands of a few global companies or TNCs. Of the world's 100 largest economies, 50 are now global corporations and *not* countries;

- Countries are increasingly interdependent because they need to trade with each other and exchange goods; the lives and actions of people in one country are increasingly linked with those of people thousands of kilometres away.

The gap between rich and poor shows that globalisation does not affect all parts of the world equally.

Figure 21 shows the proportion of internet users compared to the total population of world regions. The graph shows the percentage of internet users and how it has changed in some countries.

Figure 21
Internet users still a global enclave.

The large circle represents world population
Pie slices show regional shares of world population
Dark wedges show Internet users

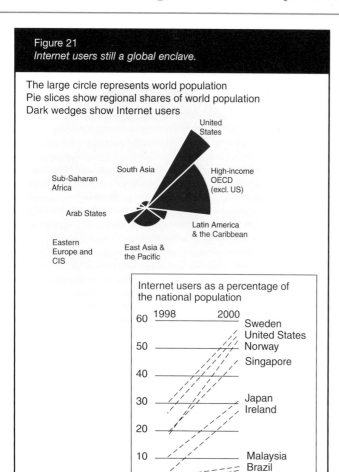

Internet users as a percentage of the national population

1998   2000
60
50   Sweden
      United States
      Norway
40   Singapore
30   Japan
      Ireland
20
10   Malaysia
      Brazil
0   South Africa
      China

activities

1 Describe the change in the pattern of internet users from the graph (Fig. 21). [4]

HIGHER TIER

2 Describe the world distribution of internet users in relation to population distribution. [6]

3 To what extent is internet use a good indicator of development? [4]

web link & extra resources

Some people benefit more than others from globalisation.

www.oneworld.org

www.belfastdec.org

www.incspotlight.org/case/index.html

Figure 22
Six situations in which globalisation affects people.

1. Global communications

It is becoming cheaper and easier to communicate. Jake in New York can phone his cousins in Cork for 3 minutes for only 35 US cents. Ten years ago, the same call would have cost $3. In 1930, when Jake's grandfather first arrived in America from Cork, it would have cost him $245 to phone home. If Kathy in the USA saves a month's wages, she can buy the most up-to-date computer. A worker in Bangladesh would have to save 8 years' wages to buy a computer. Jake and Kathy can surf the web and understand four out of every five websites – because they are written in English. However, nine out of every ten people in the world don't understand English.

2. Global advertising

Advertisements for goods like Coca-Cola and Levi's can be seen everywhere, from Belfast to Berlin to Beijing in China. In the 10 years from 1987 to 1997, Nike increased its spending on ads from $25 million to $500 million. Tiger Woods, the golfer, is paid $55,555 per day to advertise Nike products. Workers making Nike products in Indonesia earn around $1.25 per day. This is often not enough to live on.

3. Trade is increasing

Three times more goods are shipped or flown between countries now than 20 years ago. Trade can create jobs in poor countries, but people often have to work in terrible conditions. On banana, coffee and flower farms, for example, workers get very low wages, work long hours and have to use dangerous pesticides or chemicals. Some countries are missing out on trade altogether: the 48 poorest countries are being left behind and making less and less money from trade compared to richer countries.

4. Global rules

The World Trade Organisation (WTO) was set up to write the rules for trade between countries and to make sure that all countries obey them. This could be very good – fair rules would mean that everyone would win. However, the rules are written to suit rich countries. This is because rich countries can afford to pay lawyers to debate the rules, while poor countries can't. For example, Japan has 25 trade experts at the WTO, Bangladesh has one, and 29 of the poorest countries have none. There are up to ten meetings every day. Poor countries can't go to all the meetings and so important decisions are made without them.

5. Travel & tourism

Travelling abroad for holidays can be interesting and exciting. People make friends from foreign countries and may discover many interesting things about other people's lives and the global environment.

However, some people fear tourism can have a negative side. In many countries, local people look and dress in beautiful costumes and jewellery, such as the Maasai cattle farmers in Kenya. Sometimes, tourists may treat local people like objects to be photographed rather than human beings who can feel and talk.

6. The media

Television, films, radio and newspapers can help people to learn about what is happening in other places and what it would be like to live there. On the other hand, the media may also give people a false impression of other countries. In India, many people think that life for everyone in Europe and the US is exactly as it appears in the movies: rich and exciting. Meanwhile, news reports about Africa often only show images of desperate hungry people. However, most people in Africa have little money, but find many ways to make their lives happy and full, and are certainly not helpless or starving.

activities

1 Work in pairs to sort out who are the 'winners' and 'losers' from globalisation in each situation in Fig. 22. [6]

HIGHER TIER

2 Sort the following list of effects of globalisation into two columns – for and against. [6]

- The wealth created by globalisation can help to lift millions of people out of poverty.
- Trade barriers put up by MEDCs cost the economies of the LEDCs about $100 billion per year.
- The more world trade grows the wider becomes the gulf between the LEDCs and MEDCs.
- Globalisation means more links between countries and the spread of democracy and agreements on standards for the environment and human rights.
- The average American makes over 30% more than the average German, 50% more than the average UK citizen and 5500% more than the average Ethiopian and this will double in the next 150 years.
- The number of people living in poverty in many parts of the world will almost double by 2008.
- Economic growth is essential if world poverty is to be reduced.
- Wealthy MEDCs gain more than twice as much from free trade around the world while Africa actually loses.
- The spread of technology can help LEDCs to develop their industries and infrastructure.

TRANSNATIONAL CORPORATIONS (TNCS)

A transnational corporation or TNC is a global company with factories in many countries. The business of the company is usually secondary manufacturing of products which are sold worldwide. The headquarters of the company is in a MEDC, but the factories are in both MEDCs and LEDCs. The scale of TNCs means they have the capital to run very efficiently at low costs, which earn high profits to invest in research and development and in marketing their products. TNCs have grown in size and influence so that some of the larger ones make more money in a year than all the African countries put together. They control most of the world's trade and investment, produce most of the world's manufactured goods, and contribute to modern technology and scientific discoveries.

Investment is the capital or money attracted to one country from another country. TNCs often invest in LEDCs, but foreign investment is very limited in the world's poorest countries where 20% of the population live.

Relocation means the movement of one company or industry from one place to another. TNCs often move a factory from a MEDC to a LEDC.

One local example of relocation is Avalon Guitars Ltd (formerly Lowden Guitars) of Newtownards, Co. Down (see Fig. 23).

The main reasons for **not** building a new factory in Northern Ireland were:

- Being able to increase the profits without the need to invest capital in a new factory;
- Faster deliveries to customers from other countries (10 weeks instead of 16–24);
- Product range could be expanded using expertise of guitar makers abroad.

The following factors were important in the choice of location of partner factories by Avalon Guitars:

- The high levels of productivity and easy access to capital especially in Korea give it a competitive advantage;

Figure 23
Avalon Guitars

This company specialises in hand-made guitars on a limited production basis at the rate of 1500 per year at its factory in Newtownards, where it employs 32 people. It also has a distribution centre based in Texas, USA as 50% of the products are sold in North America. The increase in demand is greater than the capacity available at the factory in Newtownards and so the company decided to extend the production of acoustic guitars to the Far East. After researching 20 factories abroad, four partner factories in Korea, China, the Czech Republic and Spain were chosen to make Avalon guitars. The HQ in Newtownards will continue to be involved in the design, research & development, marketing and the quality control of the product. The local company is becoming a design-led global brand in the acoustic guitar market.

Economic Change and Development

■ Cheaper labour costs, especially in China – a typical factory worker earns US$1 per day (and will usually have food and dormitory accommodation provided);

■ High levels of investment in technology especially in Spain where skilled labour is available to make the specialist 'classical' guitars.

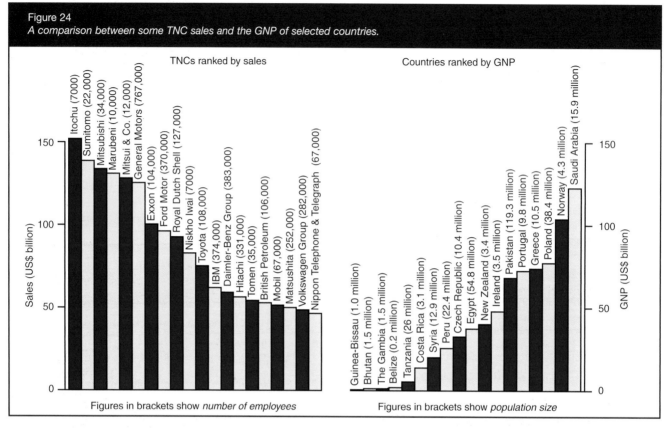

Figure 24
A comparison between some TNC sales and the GNP of selected countries.

TNCs ranked by sales

Countries ranked by GNP

Figures in brackets show *number of employees*

Figures in brackets show *population size*

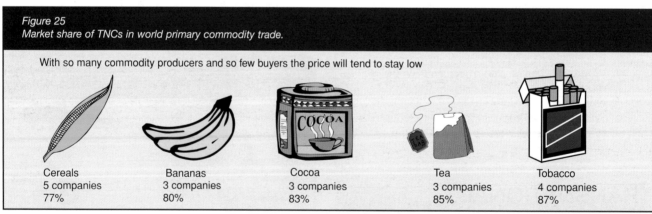

Figure 25
Market share of TNCs in world primary commodity trade.

With so many commodity producers and so few buyers the price will tend to stay low

Cereals	Bananas	Cocoa	Tea	Tobacco
5 companies	3 companies	3 companies	3 companies	4 companies
77%	80%	83%	85%	87%

Figure 26
The profits made by three TNCs in Brazil over a 20-year period.

TNC (HQ)	Profits made as % of original investment	% of profits sent back to HQ in home country
VW (Germany)	+300%	80%
Pirelli (Italy)	+350%	70%
Exxon (USA)	+1875%	40%

activities

1 Describe what the information in Figs 24, 25 and 26 shows about the importance of TNCs. [9]

HIGHER TIER

2 Explain how this might affect LEDCs. [5]

activities

1 (a) Carry out an Internet search on one TNC, then copy and complete the following table. [3]

	General location	TNC
HQ of TNC	City in MEDC	
Research and development sections	Smaller cities and suburban areas of cities in MEDCs	
Production sections	In LEDCs (low-cost locations)	

(b) For your chosen TNC find out two reasons why this TNC moved some of its production to a factory in a LEDC. [6]

Who really gains?

People are now concerned about the influence of TNCs because of their effects on people's lifestyles and especially on LEDCs. There is concern that TNCs act on behalf of their owners and shareholders who demand the highest possible profits and who live in MEDCs. The factories, farms and jobs of TNCs in a LEDC are part of the larger organisation so they depend on the decisions made at the headquarters of the company. TNCs play a major role in agriculture in LEDCs, because they control the production and export of foodstuffs such as coffee and cocoa. Some people recognise the advantages TNCs bring in helping to promote development by providing jobs and bringing new technology to LEDCs. Other people think that TNCs locate in LEDCs just to take advantage of the low wages and less strict environmental regulations, both of which help TNCs to make a huge profit.

Figure 27
Benefits and problems which TNCs can bring to LEDCs.

I am in favour of transnationals – they bring many benefits

TNCs:
- bring new investment into a country's economy.
- provide jobs, often at higher wage levels than the local average.
- provide management skills that may be lacking.
- provide research and development that can assist local development.
- international links can gain a country access to world markets.

TNCs:
- frequently import their own inputs rather than use local products.
- can damage the local environment and deplete resources.
- export profits, draining wealth from the country's economy.
- usually provide low-skilled jobs in LEDCs and retain skilled jobs in MEDCs.
- are mobile and can leave a country as quickly as they arrived.
- are powerful and can interfere in a country's political processes.
- can weaken workers' rights by looking for countries with cheaper labour.

I am against transnationals – they look after themselves rather than us

web link & extra resources

The example of Nike:

www.nikebiz.com

www.trocaire.org/campaigns/trade/tradeintro.htm

www.nikewages.org

www.cleanclothes.org/companies/nike.htm

CASE STUDY

ONE TRANSNATIONAL CORPORATION (TNC) AT THE GLOBAL SCALE

Nike is a TNC based in the USA providing about 500,000 jobs worldwide, mainly in LEDCs. It is one of the largest and most profitable shoe and clothing companies in the world, operating in over 80 countries. Nike is about selling more than trainers, it is selling a lifestyle.

As Phil Knight, its founder, said, "the mission of Nike is not to sell shoes but to enhance people's lives through sport and fitness and to keep the magic of sports alive". Nike uses sports stars to promote its famous logo worldwide and pays athletes generous amounts, e.g. Tiger Woods receives US$100 million for a 5 year period while Lleyton Hewitt is given US$15 million! The reality for many workers overseas making Nike shoes and clothing is far less rosy.

In order to develop flexibility, Nike products are manufactured by contract suppliers operating throughout Asia. The company shifts production and seeks out locations with low labour costs. The reason behind this is the competition Nike faces from other TNCs such as Adidas and Reebok. They want to keep production costs to a minimum and increase their profits. LEDCs have large supplies of cheap labour which make them attractive to TNCs such as Nike.

Nike and the Environment

Nike has made positive changes in safety measures and environmental health. In 1998 Nike announced a programme to replace petroleum-based solvents with safer water-based compounds. In 1999 an expert went into a Vietnam factory and verified that the factory had substituted less harmful chemicals in its production. The report concluded that the factory had implemented important changes that significantly reduced the exposure of workers to toxic solvents, adhesives and other chemicals.

However, there are still significant health and safety issues remaining and workers in some sections of factories face over-exposure to hazardous chemicals, and to high levels of heat and noise pollution.

Figure 28
Nike around the world.

Nike is a well-known US-based transnational corporation that specialises in manufacturing sports shoes and other sportswear. The company operates in many countries around the world and there are few countries where its products cannot be bought. The company was set up in Beaverton, Oregon, USA in the 1970s by a couple of college graduates who had an idea for a new sports shoe. The company headquarters is still in Beaverton, where most of its US employees work.

By 1998 the company had a market value of US$8 billion. Its annual sales were $6135 million and its profit was $491 million.

Today, Nike directly employs around 20,000 people worldwide. Most of these people are employed in product design, marketing and administration, working in the USA. However, there are probably another 500,000 people around the world who are employed in factories making Nike products. Most of them work in Asia for companies to which Nike subcontracts most of its manufacturing. There are factories making Nike products in Taiwan, South Korea, Hong Kong, China, Indonesia, Vietnam, the Philippines and Bangladesh. Most employees are semiskilled and unskilled assembly-line workers. They earn a small fraction of the wages of a factory worker in an MEDC. Some people argue that TNCs like Nike exploit the cheap labour in these countries. However, Nike argues that it is helping to provide jobs that are often well above the local wage rate.

Source: *GCSE Geography in Focus* by J Widdowson *et al.* (John Murray, 2001)

Figure 29
The Nike story.

The story

1964–70: A small company in Oregon, USA, called Blue Ribbon Sports (BRS), is a distributor for a make of Japanese athletic shoes.

1971: BRS changes its name to Nike and designs its own athletic footwear. Nike has a contract with another Japanese company to make the shoes.

1974: Nike products become more popular, but costs are rising in Japan. So, the Japanese manufacturers begin to subcontract shoe assembly to low cost factories in Taiwan and South Korea. Materials and pieces for the shoes still come from Japan.

1975–86: The Nike market goes global. The company opens factories in the USA, Hong Kong, Britain and Eire (Republic of Ireland). However, costs are too high and they are closed.

1983: Nike ends the manufacturing contracts with Japan because of high wages. Taiwan and South Korea become the production centres.

1985–95: As Taiwan and South Korea become more industrialised, they supply more of the materials and pieces as well as assembling the shoes. But costs rise here too. So, they continue to make specialist shoes, but the Taiwanese and South Korean companies subcontract low cost producers in Indonesia, Thailand, Malaysia, Vietnam and China to make mass-market footwear. The bigger subcontractors can make up to 30,000 pairs a day.

Source: *The World* by R. Prosser (Stanley Thornes, 1998)

Figure 30
Costs and profits of Nike.

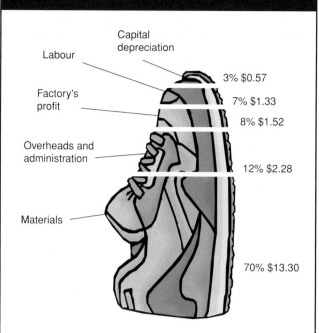

Labour — 3% $0.57

Capital depreciation — 7% $1.33

Factory's profit — 8% $1.52

Overheads and administration — 12% $2.28

Materials — 70% $13.30

NIKE TO TAKE A HIT IN LABOR REPORT

Nike has suspended a manager of its Vietnam plant in the wake of a report by a US-based labor group charging worker abuse in the athletic shoemaker's factories.

Thuyen Nguyen of Vietnam Labor Watch inspected Nike factories in Vietnam in both escorted and surprise visits.

He found violations of minimum wage and overtime laws, as well as physical mistreatment of workers, most of whom are women between the ages of 15 and 28.

'While Nike claims it is trying to monitor and enforce its code of conduct, its current approach to monitoring and enforcement is simply not working,' Nguyen say.

From Nguyen report:

• As punishment for wearing non-regulation shoes, 56 women were forced to run laps around the factory. Twelve women fainted and had to be taken to a hospital.

• Inspecting pay stubs Nguyen says he saw a pattern of paying new workers subminimum wages.

Nike spokeswoman McLain Ramsey confirms Nguyen's visit to the Ho Chi Minh City plant and says company officials are 'as distressed as he is' about his report.

The manager accused of making women run laps has been suspended, she says.

The Sam Yang plant in Vietnam is owned by a Taiwanese company which Nike subcontracts, she says.

Nike has drawn fire from worker rights groups for failing to hold its subcontracted factories in South-east Asia and China to fair labor standards. About 75% of Nike's production is done in Indonesia, China and Vietnam.

Source: *The Guardian*, 28 September 1996

Figure 31
Four views of people connected with TNCs involved in the garment industry.

A spokesperson for a TNC making clothes in factories in LEDCs

We invest heavily in LEDCs and pay higher wages than the local average: we often have to contract production out to local factories in order to produce products quickly and to keep production costs low. This is in order to keep at the forefront of competition with other TNCs in the same business.

A development worker in Australia

TNCs in the garment industry need to be flexible and so they move production very quickly from one factory (and one country) to another. This means they sub-contract work out to local factories in LEDCs who are eager to make a quick profit and who often ignore the workers' rights as their trade unions may be weak.

A worker in a factory in Thailand supplying clothes to a TNC

We have plenty of work, making clothes for the TNC The work is hard and the hours are long – we can even work for up to 110 hours per week. We are able to earn much higher wages than in other local jobs and so our families can have a much higher standard of living.

A trade unionist in a textile factory in Indonesia

FACTORY CLOSED

When we lost the orders for clothes for the TNC, the factory was quickly closed because we only had one TNC as the customer for our products The local factory owner disappeared and so no-one was made responsible for paying our workers their backpay.

Figure 32
Reebok challenges Nike.

REEBOK LEADS CHILD LABOUR PURGE

By Roger Lowe

Reebok yesterday called on its rival Nike, the market leader in trainers, to join it in a bid to end child labour and improve working conditions at their Asian factories.

Paul Fireman, the chairman and chief executive, has written to his opposite number at Nike, Phil Knight, proposing joint monitoring of factory conditions.

Reebok and Nike – who together sell 60 per cent of branded trainers – have come under pressure in Britain and America in campaigns against child labour and exploitation of Asian workers.

Reebok has taken a public stand on social issues, launching the Reebok Human Rights Award in 1988 and adopting a Human Rights Production Standard in 1992. Earlier this year, after accusations of child labour in Pakistani factories, it announced a new factory which would be monitored to ensure no children were employed.

Most trainers are made in Indonesia, the Philippines, China and other Asian countries, although production has moved from Taiwan and South Korea to lower-cost countries where trade unions are often banned.

Christian Aid says only £1.20 from the price of a £50 pair typically goes to workers who made the shoes, although huge sums are spent on marketing. It also notes that Chinese workers would have to work nine hours a day, six days a week for 15 centuries to earn the £929,113 paid to Nike boss Phil Knight last year.

The charity, in conjunction with the Fairtrade Foundation, has developed a code of conduct setting out minimum conditions and the need for independent monitoring.

But charities have also warned that a ban on child labour could be counter-productive. Oxfam argues that in very poor countries every family member has to work if they are to survive. It cities examples of Bangladeshi children forced into prostitution when factory owners stopped employing them after fears of a boycott in 1993.

Source: The Guardian, 28 September 1996

Figure 33
Nike factories and GNP.

Country	GNP per person (US$)	Number of Nike factories
China	840	59
India	450	21
Indonesia	570	32
France	24090	1
Germany	25120	1
Mexico	5070	35
Portugal	11120	23
Japan	35620	7
Thailand	2000	61
UK	24430	4

Source: www.nikebiz.com and World development indicators database, World Bank.

activities

Answer these questions using Figs 28–32.

1 Read The Nike story (Fig. 29).

(a) On a world map, mark the countries which have produced Nike products since 1960. [4]

(b) Describe how their locations have changed over time. [4]

2 Using the information provided, explain why Nike has moved production from MEDCs to LEDCs. [6]

3 List the advantages and disadvantages Nike has brought to people in LEDCs [9]

4 Draw a spider diagram to show the advantages and disadvantages NIKE has brought to the environment in LEDCs. [6]

HIGHER TIER

5 (a) Outline the attitudes to child labour in both LEDCs and MEDCs. [6]

(b) Do you think child labour in TNCs is justified? Explain your answer [4]

6 Study Fig. 33, this shows the number of Nike factories in selected countries and the wealth of those countries (as measured by their GNP). Answer these questions.

(a) Draw a scatter graph (using ICT) and mark a best-fit line to show the relationship between the number of factories and GNP. [6]

(b) Describe the relationship shown and what it indicates about the influence of Nike as a TNC. [6]

UNIT THREE
Sustainable Development Strategies

DIFFERENCES IN DEVELOPMENT BETWEEN MEDCS AND LEDCS

Measuring development

Geographers are interested in differences in levels of development and rates of growth between places across the world and within countries. There are many ways of measuring a country's development and each indicator of development or development index has its own advantages and disadvantages. Traditionally, development was measured in economic terms alone – usually in GNP (Gross National Product) per person. Based on wealth, countries can be divided into the economically more-developed countries which include the rich industrialised countries of the 'North', for example, Canada, the United Kingdom and Germany, and the economically less-developed countries of the 'South' such as Sierra Leone, Nepal and Burkina Faso.

Figure 34 shows the North–South divide, which illustrates the differences in economic development between the rich 'North' and the poor 'South'.

However, measuring development by wealth alone identifies a number of countries which, while economically wealthy, could not be described as developed. As GNP is based on averages it hides the existence of widespread poverty in rich countries, and great wealth in some poorer countries. It reduces the measure of development to one variable – economics. It is therefore important to consider other factors when measuring development, especially the indicators of the social welfare of a population.

The indicators of development can be subdivided into social indicators which relate to people's well-being, that is they measure *human welfare*, and economic indicators which are related to wealth. Figure 35 shows some socio-economic indicators of development for a range of countries. The table suggests that there are clear links between wealth and a range of social factors. Economically more developed countries have lower birth and infant mortality rates, a longer life expectancy, a slower natural population increase, higher adult literacy rates and fewer people per doctor than countries which are economically less developed.

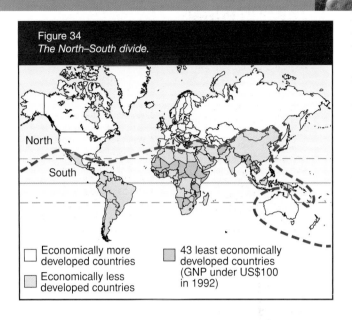

Figure 34
The North–South divide.

North

South

☐ Economically more developed countries

☐ Economically less developed countries

▨ 43 least economically developed countries (GNP under US$100 in 1992)

Problems with indicators

One major problem with social indicators is that the information is often obtained from a census or household survey, which may not be accurate. In addition, some social indicators are also directly related to the wealth of a country. For example, the more wealthy a country is, the more TVs and computers it has per 1000 people, the lower its infant mortality rate, the fewer of its population per doctor due to better health care and the higher adult literacy rate due to better education.

The use of separate indicators can also disguise variations in development. For example, when comparing GNP between countries, those with high levels of subsistence agriculture will rate poorly, even though the majority of their population will be well fed. This is because subsistence agriculture does not contribute to the country's earnings. For this reason some composite measures have been developed by the United Nations Development Programme (UNDP), which combines a range of indicators. Three of these measures are:

- The Human Development Index;
- The Human Poverty Index;
- The Gender-related Development Index.

The Human Development Index (HDI)

This index is expressed as a figure between 0 and 1 (see Fig. 36). Countries can be ranked according to

Figure 35
Socio-economic indicators of development.

Country	Human Development Rank	GNP per capita (US$) 1998	% of Population below the income poverty line	Life expectancy (years) 1995–2000	Under 5 mortality rate (per 1000 live births) 1998	Annual population growth rate (%) 1975–1998	Urban population as % of total population	Adult literacy rate (%)	Number of pupils to one Primary teacher 1995	Population without access to safe water 1990–1998	Doctors per 100,000 people 1992–1995	TVs per 1000 people 1996–1998	Computer per 1000 people 1996–1998
Canada	1	19,170	5.9 dd	79.0	6	1.2	76.9	99.0	16	X	221	715	330
U. Kingdom	10	21,410	13.1 dd	77.2	6	0.2	89.4	99.0	19	X	164	645	263
Germany	14	26,830	11.5 dd	77.2	5	0.2	87.1	99.0	18	X	283	580	305
Ireland	18	18,710	36.5 dd	76.2	7	0.6	58.1	99.0	23	X	167	456	272
Italy	19	20,090	2.0 dd	78.2	6	0.1	66.8	98.3	11	X	X	486	173
Russian Federation	62	2,260	X	66.6	25	0.4	77.0	99.5	20	X	380	420	41
Brazil	74	4,630	5.1 dg	66.8	42	1.9	80.2	84.5	23	24	134	316	30
Thailand	76	2,160	28.2 dg	68.8	44	1.7	20.9	68.8	20	19	11	236	22
Philippines	77	1,050	18.7 dg	68.3	44	2.3	56.9	68.3	35	15	11	108	15
China	99	750	X	69.8	47	1.3	32.7	69.8	24	33	115	272	9
South Africa	103	3,310	11.5 dg	54.7	83	2.0	49.9	84.6	37	13	59	125	46
Lesotho	127	570	43.1 dg	56.0	136	2.4	26.4	82.4	49	38	5	24	X
India	128	440	44.2 dg	62.6	105	2.0	27.7	55.7	63	19	48	69	3
Nepal	144	210	37.7 dg	57.3	100	2.6	11.2	39.2	39	29	5	4	X
Bangladesh	146	350	29.1 dg	58.1	106	2.1	20.0	40.1	X	5	18	7	X
Burkina Fasa	172	240	61.2 dg	44.4	165	2.7	17.4	22.2	58	58	X	6	1
Sierra Leone	174	140	57.0 dg	37.2	316	1.9	35.3	33.3	X	66	X	26	X

Key Indicators of Development

Adult Literacy Rate The percentage of people aged 15 and over in a country who can read and write.

Food Production per capita The average annual quantity of food produced per capita.

GDP Gross Domestic Product per head. The value of all the goods and services produced by a country in a year divided by the country's population.

GNP Gross National Product per head (US$ per capita per year). The total value of the goods and services produced in that country in a year divided by the country's population.

Infant Mortality The number of deaths of infants under 1 year per every 1000 live births. Can be expressed in relation to under 5 year olds.

Life Expectancy The average age to which people are expected to live in a country.

Population Growth Rate The percentage by which a country's population is growing.

Urban Population % The percentage of a country's population who live in urban areas.

Human Development Rank Country's development rank according to HDI score.

Figure 36
The Human Development Index across the world.

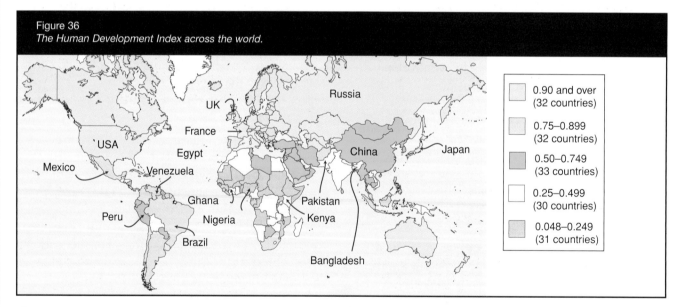

	0.90 and over (32 countries)
	0.75–0.899 (32 countries)
	0.50–0.749 (33 countries)
	0.25–0.499 (30 countries)
	0.048–0.249 (31 countries)

their HDI score. The closer the score is to 1 the more developed the country is. The index combines measures of health, wealth and education as follows:

- Life expectancy at birth measured in years. This measure has a two-thirds weighting.

- Literacy rate for adults over 15 years and the combined primary, secondary school and third-level education enrolment. This was weighted at one third.

- GDP per capita measured in US$.

Countries scoring low values, that is, the less-developed countries, include Sierra Leone and Burkino Faso. Those which are highly developed, with high scores, include Canada and Norway.

Some geographers feel the HDI still puts too much emphasis on wealth and suggest other measures should be incorporated, for example freedom of speech.

activities

1 Study Fig. 36, showing the HDI across the world.

(a) Describe the distribution of the 31 least-developed countries according to the HDI.

(b) Name a country which is located in the 'North' in Fig. 34 but which fails to achieve a score in the top category for development (>0.90), according to the HDI in Fig. 36.

(c) Name a country which appears in the less-developed 'South' in Fig. 34 but which has a relatively high HDI (0.75–0.89), in Fig. 36.

(d) As the North–South divide is based on GNP per capita, what does this tell us about using GNP alone as an indicator of development?

[8]

2 Name five indicators of development and describe how they are used to describe variations in development between countries.

[10]

3 Thinking Skills Activity: Development.

1. birth rate	2. most people in primary industry
3. Asia	4. Europe
5. high life expectancy	6. subsistence agriculture
7. low GNP	8. natural increase
9. high birth rate	10. low infant mortality
11. MEDCs	12. South America
13. little trade	14. migration
15. agribusiness	16. high GNP
17. death rate	18. high death rate
19. LEDCs	20. Africa

Task 1

Each of the numbers in the sets of four relates to the topic development. Can you work out with your partner which is the *Odd One Out* and what connects the other three?

Set A	14	17	1	8
Set B	17	11	10	16
Set C	7	19	9	5
Set D	2	15	13	6
Set E	20	3	4	12

Task 2

Still with your partner, can you find *one more* from the list to add to each of the sets above so that *four* items have things in common, but the *Odd One Out* remains the same? Think about why you have chosen each one.

continued

Task 3

Now it's your turn to design some sets to try out on your partner! Choose three numbers that you think have something in common with each other and one that you think has nothing to do with the other two. Get your partner to find the *Odd One Out*, then do one of theirs. Try a few each, but remember to be reasonable.

Task 4

Can you organise all the words into groups? You are allowed to create between three and six groups, and each group must be given a descriptive heading that unites the words in the group. Try not to have any left over. Be prepared to rethink as you go along.

Source: *Thinking Through Geography Material*, N.I. Education Board (1999).

HIGHER TIER

4 (a) Use the internet to obtain the latest GNP and HDI figures for the UK, the Russian Federation, Germany, Sri Lanka, Columbia, South Africa, Botswana, Saudi Arabia, China and Sierra Leone. Rank all of the countries according to their GNP per capita, the country with the highest GNP should be ranked 1 and the lowest 10. Now repeat for the HDI, ranking the country that performs best as 1 and the country that performs worst 12.

(b) Plot the results on a scattergraph:

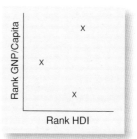

(c) Describe and account for the resulting pattern.

(d) If you were designing an index to show development what factors would you consider?

[25]

5 Read the information about the development diamond below and then study the diamonds for the countries overleaf (Fig. 37).

Development Diamond

The development diamond portrays four selected socio-economic indicators for a given country, compared to the corresponding averages for the income group to which the country belongs. The standard for the country is shown by the black diamond, while the actual conditions in the country are shown in colour. Countries are better than the average for their group when they are positioned beyond the black diamond, and they are less well developed when their score can be plotted inside the diamond. The group averages for gross primary enrollment, access to safe water, and GNI per capita are weighted by the individual country's population. The group average for life expectancy is weighted by the number of live births. GNI is gross national income, a measure of domestic and foreign income.

What does the diamond suggest about development in Sierra Leone? [4]

HIGHER TIER

6 (a) Compare the level of development of each of the countries compared with others from the same income group.

(b) Which indicator would you consider to be the most accurate for describing the level of development in Oman? Give reasons for your answer.

[12]

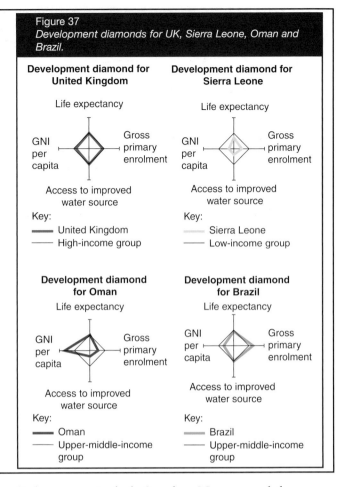

Figure 37
Development diamonds for UK, Sierra Leone, Oman and Brazil.

web link & extra resources

www.ciesin.columbia.edu/IC/IC-partners.html

www.worldbank.org

www.worldbank.org/data/countrydata/aag.htm

www.un.org/esa

The Human Poverty Index (HPI)

This index measures deprivation. It measures three factors:

- Early death rates – the probability of not living to the age of 40.

- Adult illiteracy – the proportion of adults unable to read or write.

- Living standards – as measured by the percentage of the population not using improved water sources and the percentage of children under 5 years who are underweight.

Countries with high scores where there is much poverty include Burkino Faso, Mozambique, Sierra Leone and Burundi. Countries with low scores with little poverty include Sweden, Norway and the Netherlands.

Gender-related Development Index

This index measures the same characteristics as the HDI but this time the facts are considered on a gender basis that is looking at inequalities between men and women.

The scores are largely similar to the HDI values with some variation in the middle development group.

In conclusion it can be seen that there are vast differences between countries in terms of development. However assessed, the gap is ever widening and cannot be sustained environmentally, politically or economically. Gustave Speth of the UNDP suggests:

"Sustainable human development is development that not only generates economic growth but distributes its benefits equitably; that regenerates the environment rather than destroying it; that empowers people rather than marginalizing them. It gives priority to the poor, enlarging their choices and opportunities, and provides for their participation in decisions affecting them. It is development that is pro-poor, pro-nature, pro-jobs, pro-democracy, pro-women and pro-children."

Figure 38
Differences in development between MEDCs and LEDCs.

	MEDCs	LEDCs
Gross National Product (GNP)	The majority have a GNP of more than US$5000 per year. MEDCs have 83% of the world's income.	The majority generate less than US$2000 per year. Collectively, LEDCs have only 17% of the world's income.
	High percentage of population is above poverty line; that is, they have an income of more than US$14.4 per day.	High percentage of population is below the poverty line; that is, they have an income of less than US$1 per day.
Life expectancy	Over 75 years.	Under 60 years.
Population and population growth	25% of world population.	75% world population
	Relatively slow growth, for example 25% of population doubles in 80 years.	Fast growth, for example 75% of population doubles in 30 years.
	Effective family planning.	Little or no family planning.
Disposable income spent on consumer goods	High – large numbers of consumer durables per 1000 people due to large disposable income.	Low – few consumer durables per 1000 people.
Health	Good – relatively few people per doctor.	Poor – large numbers of people per doctor.
	Account for 94% world health expenditure.	Account for only 6% of the world health expenditure.
	Well-equipped hospitals.	Inadequate hospital provision and medication.
Education	Account for 89% of the world's education spending.	Account for 11% of world's education spending.
	The majority have full-time secondary education.	Few have formal education opportunities.
	Good teacher pupil ratio in schools (Primary).	Poor pupil teacher ratios.
	High adult literacy rates.	Low adult literacy rates.
	No gender bias in educational opportunities.	Females disadvantaged in educational opportunities.
Employment structure	Large % of population involved in secondary and tertiary industry.	High % of population involved in primary industry.
	75% of world's manufacturing industry.	25% of world's manufacturing industry – much of it TNC owned.
Levels of technology/ mechanisation	Highly mechanised.	Mostly manual labour and animal power.
	Large investment in research and development.	Little native investment in research and development.
	92% of world's industry.	8% of world's industry.
Diet/access to clean water	Balanced diet.	Much malnutrition.
	High animal protein diet.	Low protein diet.
	70% of world's food grains.	30% of world's food grains.
	Majority of population have access to clean water.	Many people do not have access to clean water.
Energy	High levels of consumption, use 75% of world energy.	Lower consumption rates, currently 25% of world energy.
Communications	Good communications infrastructure: roads, railways and airports.	Communication infrastructure focused on urban areas. Limited elsewhere.
Exports	82% of world's export earnings.	18% of world's export earnings.
	Mostly manufactured goods.	Based on primary products and unprocessed raw materials.

APPROPRIATE TECHNOLOGY AND ECONOMIC DEVELOPMENT

Technology is the application of scientific, computerised or automated methods to work or industrial processes using modern machinery or equipment.

Appropriate technology is technology that uses the skills and suits the needs and level of wealth of local people.

Use of appropriate technology is more likely to encourage economic development in a sustainable manner. **Sustainable development** means "meeting the needs of the present, without compromising the ability of future generations to meet their own needs" (G. H. Brundtland, "Our common future", 1987).

Appropriate technology contributes to a more sustainable way of life for all regardless of whether they live in MEDCs or LEDCs. In MEDCs the technology is likely to be hi-tech. In LEDCs alternative forms to technology should be used, for example labour-intensive projects, technology appropriate to skills of locals, low-cost schemes which people can afford, and that include use of local craft skills in a way which does not harm the environment.

Sustainable development aims at maximising the economic, environmental and social benefits of development while minimising the economic, environmental and social drawbacks. There is often conflict between economic development and the protection of the environment. Traditionally, economic development was seen in terms of large-scale industrial projects, but sustainable development is more likely to focus on small-scale projects that are for the well-being of local people and have the minimum negative impact on the environment.

Sustainable development should therefore improve:

- Quality of life;
- Standard of living.

This can be done by:

- Using natural resources in a way that does not damage the environment;
- Encouraging economic development that the country can afford and so avoid debt;
- Developing appropriate technology, that is technology suited to the skills, wealth and needs of local people, which can be handed down to future generations.

In August 2002 the second World Summit on Sustainable Development (WSSD) took place in Johannesburg, South Africa. It was an opportunity to

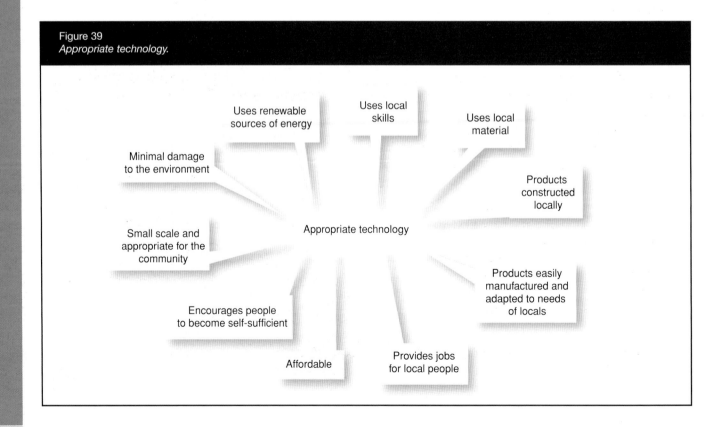

Figure 39
Appropriate technology.

Uses renewable sources of energy

Uses local skills

Uses local material

Minimal damage to the environment

Products constructed locally

Appropriate technology

Small scale and appropriate for the community

Products easily manufactured and adapted to needs of locals

Encourages people to become self-sufficient

Provides jobs for local people

Affordable

review how far the sustainable developments planned at the first Earth Summit 10 years before in Rio de Janeiro had progressed.

The Johannesburg summit focused on sustainable development in five areas, together known as WEHAB. These are:

- Water and Sanitation;
- Energy;
- Health;
- Agriculture;
- Biodiversity.

Although there have been improvements in some of these areas in the 10 years since the Rio summit, there is still room for more improvement. However, the fact that more than 22,000 people attended the meeting from 198 nations is encouraging, as it means more governments and people will be encouraged to consider sustainable development projects.

CASE STUDY

CASE STUDY ON ONE SUSTAINABLE DEVELOPMENT PROJECT: FISHING IN SOUTH-WEST INDIA

Along the shores of Kerala and Tamil Nadu, which are the most southern parts of India, local villagers depend on fishing for their income and nutrition. This area is one of the most densely populated areas in the world with 135 people/km^2. Of the working population, 70% are involved in fishing directly and 21% in fishing-related activities. These are some of the poorest people in India and they have a literacy rate of 31%, which is much lower than the literacy rate of 64% elsewhere in the Tamil Nadu state.

They carry out their fishing from canoes made from hollowed out logs and rafts called kattumarams made from tied logs. A unique feature of this raft is that it can be transformed into bigger or smaller units by adding or removing a log. The rafts are well designed to cope with the fierce south-west monsoon surf conditions.

Two problems decreased the income from fishing for the villagers:

My name is Anbazhagan. I am 65 years old and I'm the fifth generation of my family to fish these waters. Suddenly I have no work … it is because of all the deep water mechanised boat trawling. Politicians come and promise everything but they never return.

- Owing to widespread deforestation in the area there is a shortage of suitable tree trunks and light wood which is used to make the canoes and rafts. This has caused the traditional boats to become very expensive to make.

- The total catch decreased significantly between the mid 1970s and mid 1980s. It was felt that this was due to the increased use of trawlers fishing for prawns which disturbed the traditional fishing grounds.

Figure 40
A traditional southern Indian canoe.

continues

continues

Figure 41
Traditional kattumaram.

Figure 42
Fibreglass boats. Below: Kanyakumar, the southernmmost point on the Indian subcontinent.

In an attempt to address this problem the local fishermen began to use advanced technology to compete with the trawlers. They fixed outboard motors to their boats so they could travel further and avoid the trawlers paths, and also they could carry more nets. This strategy did not help the farmers as they lacked information and it was expensive. The villagers did not have access to enough money to develop and run these types of boats.

The fishermen then worked in consultation with European agencies on a number of development projects. Together they were able to devise ways of using the technology in a way more suited to the needs of the fishermen. The fishermen needed boats that were:

- unsinkable;

- able to carry engines;

- light;

- easy to launch from surf-beaten beaches;

- able to last 7–10 years;

- more comfortable and able to carry more than boats powered by sail and oars.

A solution was found in the design of a new boat.

The new boat followed the traditional centuries old designs but it was built in a different way. This preserved the uniqueness of the traditional boats in the region. The new building technology involved stitch-and-glue. The methods used the carpentry skills of traditional boat building and therefore was suitable for the local fishermen. Marine plywood and fibreglass were used as the building materials instead of the traditional tree-trunks as this reduces the cutting down of trees that are already under threat. The new boat is easily adapted to many craft designs and therefore uses the local peoples' skills while using technology which the local industries can easily adapt to. The boats have gained in popularity and the local fishermen, in consultation with European agencies, have extended the range of designs. New job opportunities have resulted in the building of the boats and by the end of 1995 there were about 5000 plywood boats in operation in the area. Between 2000 and 2002 two-thirds of all boat-owning fishermen changed from the wooden kattumarams to plastic 'fibre' boats and boats made of plywood. The transformation to appropriate technology played a vital role in the development process in this area of southern India.

continues

activities

1 (a) Copy and complete the table as shown below:

Appropriate technology	How used in the scheme for making fishing boats, SW India

(b) Select and list the statements from those given below which are true of appropriate technology in a LEDC in the left column of the table.

(c) Write a description in the right column to show how each is used in the scheme for the production of fishing boats in SW India.

[20]

Statements

■ Causes water and air pollution to the environment

■ Involves large-scale industries which require a lot of investment

■ Uses the skills of local people

■ Uses expensive imported resources

■ Employs people with a high level of education

■ Uses large amounts of resources

■ Enables people to take control of their work

■ Manufactured goods are only for export overseas

■ Requires much capital which is borrowed from foreign companies or banks

■ Is a sustainable form of development as it does little damage to the environment

■ Requires relatively few workers

■ Promotes development at affordable levels for the people

■ Allows people to have a better quality of life and standard of living

■ Uses technology which people can understand and easily use

■ Does not require large investments of capital

■ Uses local resources

■ Helps people to become self-sufficient

ICT

2 Write a newspaper report for an aid agency's newsletter explaining how appropriate technology was used to help the fishermen secure their income. [10]

3 Create a learning map to show how the development project is likely to ensure economic, environmental and social progress. [10]

4 State fully one positive and one negative impact of this boat-building project. [6]

HIGHER TIER

5 State fully two ways in which this project could be described as an example of sustainable development. [6]

web link & extra resources

www.sustainable-development.gov.uk

www.johannesburgsummit.org

WORLD TRADE

web link & extra resources

www.worldbank.org/wbiep/trade

www.wto.org

www.trocaire.org/campaigns/

www.cafod.org

www.ictsd.org

www.worldtrade.org

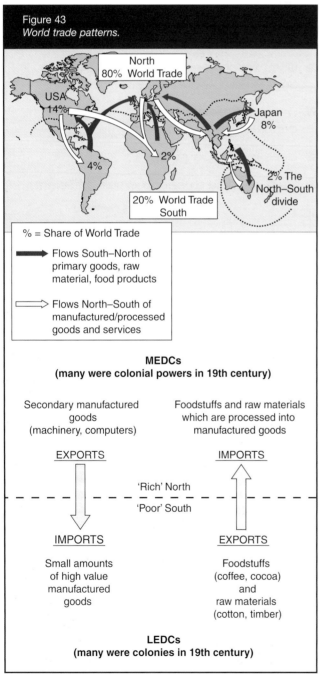

Figure 43
World trade patterns.

North
80% World Trade

USA
14%

Japan
8%

4%

2%

2% The
North–South
divide

20% World Trade
South

% = Share of World Trade

➡ Flows South–North of
primary goods, raw
material, food products

⇨ Flows North–South of
manufactured/processed
goods and services

MEDCs
(many were colonial powers in 19th century)

Secondary manufactured
goods
(machinery, computers)

Foodstuffs and raw materials
which are processed into
manufactured goods

EXPORTS

IMPORTS

'Rich' North

'Poor' South

IMPORTS

EXPORTS

Small amounts
of high value
manufactured
goods

Foodstuffs
(coffee, cocoa)
and
raw materials
(cotton, timber)

LEDCs
(many were colonies in 19th century)

Trade is the buying and selling of goods and services between one country and another. Trade takes place because no single country can provide everything that its people need or want so it must exchange goods and services with other countries. This shows the **interdependence** of countries involved in world trade. Imports are the goods bought from other countries, and to pay for these goods a country sells goods and services as exports. Imports and exports include such items as raw materials, energy resources and manufactured goods.

A **trade surplus** occurs when a country earns more money from the goods it exports than it needs to spend on buying imports. This will help the country to become richer and the money can be invested in projects which help the country to develop and improve the people's quality of life. A **trade deficit** occurs when a country has to spend more on imports than it earns from exports. This will keep the country poor so that it cannot afford to invest in industries or improve health care or infrastructure and it may have to borrow money and it is likely to get into debt.

Trade is becoming more competitive as all countries want to increase their amount of trade and to be less dependent on imports. Countries that trade with each other are said to be interdependent.

The trade patterns of MEDCs

1 The share of world trade by MEDCs is increasing and MEDCs control 80% of the world's trade; most of this trade is with other MEDCs. The European Union and the USA control over half the world's trade.

2 Many of the MEDCs have a trade surplus and some of them belong to trading blocs such as the European Union which try to increase the trade of their members and not trade as much with non-members, which are often LEDCs.

3 MEDCs export processed manufactured goods such as machinery and engineering products which fetch steady, high prices when exported to LEDCs.

4 Many MEDCs put up trade barriers against other countries so that jobs and industries in these MEDCs will be protected against cheaper imported goods from LEDCs. They do this by putting taxes or tariffs on imports and setting quotas on the amount of imports. It is estimated that trade barriers in the rich 'North' cost LEDCs $700 billion per year.

Figure 44
The commodities and relative price of exports of LEDCs.

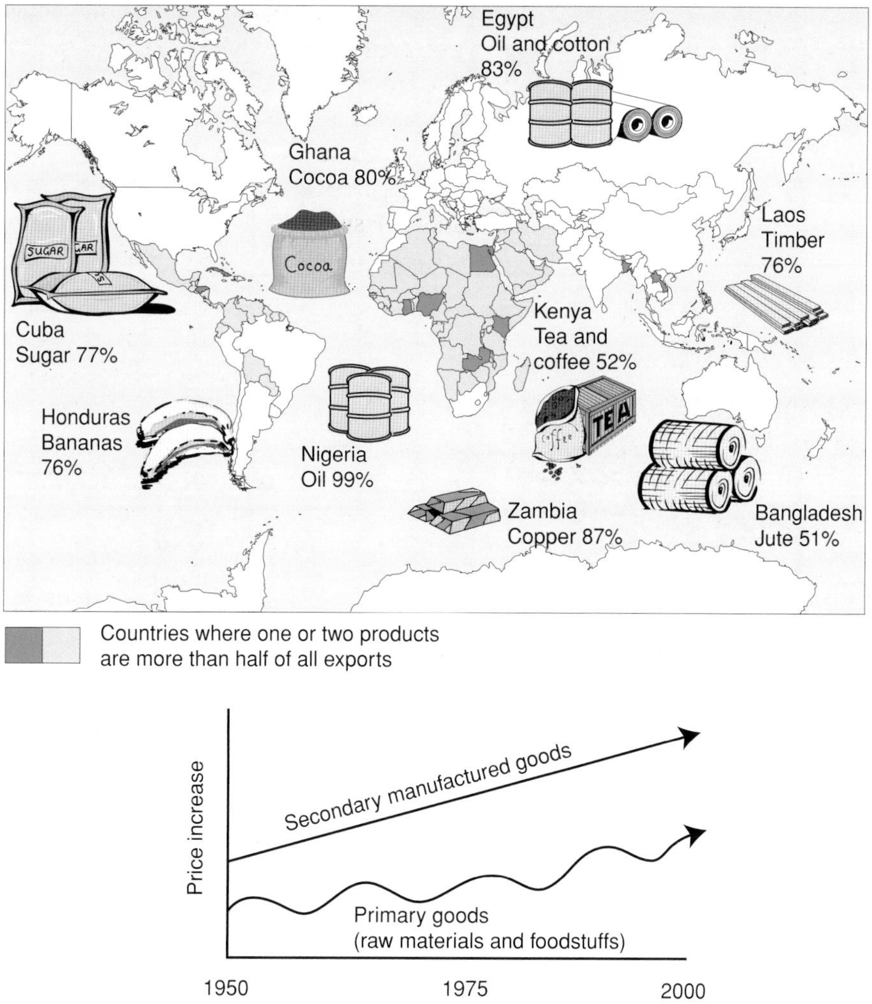

Egypt
Oil and cotton
83%
Ghana
Cocoa 80%
Cocoa
SUGAR
Cuba
Sugar 77%
Laos
Timber
76%
Kenya
Tea and
coffee 52%
TEA
Honduras
Bananas
76%
Nigeria
Oil 99%
Zambia
Copper 87%
Bangladesh
Jute 51%
Countries where one or two products
are more than half of all exports
Price increase
Secondary manufactured goods
Primary goods
(raw materials and foodstuffs)
1950
1975
2000

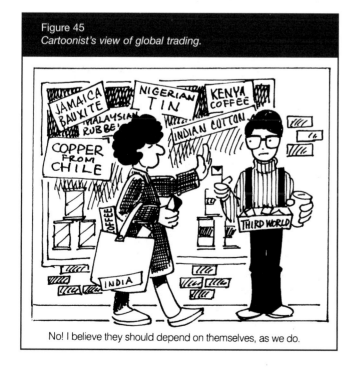

Figure 45
Cartoonist's view of global trading.

JAMAICA BAUXITE
NIGERIAN TIN
KENYA COFFEE
MALAYSIAN RUBBER
INDIAN COTTON
COPPER FROM CHILE
COFFEE
THIRD WORLD
INDIA

No! I believe they should depend on themselves, as we do.

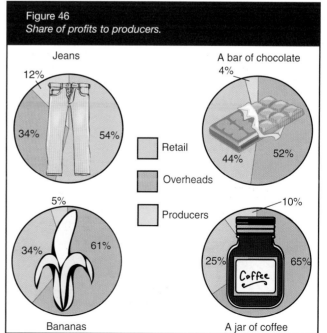

Figure 46
Share of profits to producers.

Jeans
12%
34%
54%
A bar of chocolate
4%
44%
52%
Retail
Overheads
Producers
5%
34%
61%
Bananas
10%
25%
65%
Coffee
A jar of coffee

Figure 47
Five trade traps for LEDCs.

Why don't LEDCs grow crops that will sell abroad? They can earn a good price for them.

78% of our foreign cash comes from cocoa

48% of our foreign earnings come from selling bananas

Most do sell their crops abroad. This makes them very dependent on the world prices for their goods which fluctuate from year to year.

Over the years our income from export crops has gone down. But the cost of manufactured goods we import has gone up

In 1960, one 5-tonne container of tea was enough to buy a tractor. In 2000, it took fifteen 5-tonne containers

More money is earned by processing the goods, but this is done by MEDCs

Why don't LEDCs process their own goods?

Tariffs in many MEDCs make this difficult.

We levy no duty on imports of cocoa beans

But we charge 30% duty if they want to sell us cocoa powder

...and 35% if it's chocolate bars

Surely their crops are needed by many countries? Why don't they get a good price?

Look Buddy, when the price of my coffee goes up I drink less!

Few of their products are really essential.

Why don't they stop growing for an overseas market and produce crops they need themselves?

Because they need money to pay off the debts ...both of individual farmers, and of governments.

We can't switch to growing food for the family even if the price of cotton goes down

For we've had to borrow to pay for the fertilizer

If we don't come up with the cash they'll take our land

Our balance of payments problem forced us to take overseas loans ...

...which have to be repaid

and the foreign exchange can come from exporting crops

• So, it is risky for LEDCs to be based on cash crops.

• Foreign exchange benefits only the small elite.

• Once countries and individuals are in debt it is difficult to stop growing cash crops.

The trade patterns of LEDCs

1 LEDCs have a decreasing share of world trade and control only about 20%. Only the Newly Industrialised Countries (NICs) have a larger amount of trade with MEDCs – examples are Korea, Taiwan and Brazil. These countries have rapidly increased their exports of secondary manufactured goods to MEDCs.

2 Most LEDCs have a trade deficit and do not belong to trading blocs.

3 Many LEDCs export primary unprocessed products or raw materials and foodstuffs which are sold at low and fluctuating prices. A total of 50 LEDCs depend on only three products for half their export earnings – coffee, cotton and cocoa.

4 The trade gap between MEDCs and LEDCs is becoming wider as trade becomes more imbalanced.

Fair trade

web link & extra resources

www.maketradefair.com

www.fairtradefederation.com

www.oxfam.org.uk

www.tearfund.org

www.fairtrade.org

www.craigavon.gov.uk/environment/eSustainability.asp

www.co-op.co.uk

www.trocaire.org/campaigns

www.divinechocolate.com

www.dubble.co.uk

activities

Complete the following using Figs 43–47.

1 Describe how the type of goods traded is different for LEDCs and MEDCs. [6]

2 How do trade barriers set up by MEDCs prevent LEDCs from benefiting from trade? [4]

3 Describe each of the trade traps for LEDCs shown in Fig. 47. [10]

HIGHER TIER

4 Suggest why it is a problem for LEDCs to rely on only one or two primary products to export. [5]

5 (a) Explain why the trade gap is becoming wider.

  (b) How will this affect people in LEDCs? [5]

6 'The pattern of world trade is unfair.' Justify this statement. [5]

7 Investigate the role of the World Trade Organisation (WTO). To what extent is the WTO biased towards the nations of the rich 'North'? [6]

Figure 48
Fair trade products and the Fairtrade logo.

Fair trade means that people who make or grow a product are paid fairly for their work. Producers get paid directly at fair prices, cutting out the middlemen who would have taken most profit. Fair trade sales are increasing and now account for $400 million worldwide each year. Some 4.5 million small-scale farmers already benefit from fair trade.

Figure 49
How the coffee trade affects you.

Below is a breakdown of where your money goes when you buy a cup of coffee in a typical coffee-house or restaurant. The figures are based on the average price of a cup of coffee in the UK.

Farmer	1.5p	
Agent	1p	The producing country's
Haulage	1p	total profit is a mere 11p
Marketing and drying	7.5p	
Export cost	12p	
Packaging and marketing	25p	
Roasting and processing	62p	
Retailers	60p of which 30p is profit	
TOTAL COST :	**£1.70**	

activities

Using ICT and Fig. 49.

1. Use ICT to graph and display the information on the breakdown of the price of a cup of coffee in Fig. 49. [4]

2. Describe the main features of your bar graph or pie chart. [6]

3. Suggest why the producing country receives only 11p of the total cost of a cup of coffee. [4]

4. State the percentage of the price paid which goes to the retailers. [2]

5. Explain how the retailers might justify their large profit margins. [4]

Figure 50
Divine Plain Chocolate wrapper.

Figure 51
Three stories of people who are benefiting from Fair trade.

Celestina from Tanzania

We get a higher price when we sell our coffee for 'Cafedirect' (a fair trade coffee). This means that our co-operative has been able to pay a doctor who will give treatment to our members.

(A co-operative is a group of people who have formed a business together. The members share the profits and benefits.)

The price difference has meant that I can afford more food for my family and send my children to school properly equipped with books for the first time.

Cafedirect stand by the agreed price for our coffee even when the international price of coffee falls on the world market.

Luis from Ecuador

My neighbour sells his bananas through a middleman to the world market and he gets $1 for a 40lb box of fruit. I sell my bananas to La Guelpa collective which is part of the fair trade market and I get $2.50 for every 40lb box of fruit. This means I can have a much better standard of living.

The collective has invested its money into the village and farms. There is now clean water in all of the communities and we are trying to improve the health care and build more schools.

José from the Dominican Republic

My wife and I work a small farm, growing cocoa, and together with other farmers in the local farmers' cooperative (called CONACADO) we process, market and transport the cocoa. CONACADO sells the cocoa to UK fair trade chocolate, guaranteeing a better price for us all and enabling us to strengthen the cooperative. All my five children went to school and some to further training, although I cannot read or write.

Fair trade in Coffee

The problem

Many coffee farmers receive market payments that are lower than the costs of coffee production, which keeps people in a cycle of poverty and debt. The world price for raw coffee has dropped by over 50% since 2000. When the market price of coffee drops below the producers' costs, people who produce coffee are working longer and harder for less money. The intensive production of coffee can cause other difficulties such as the loss of trees through **deforestation** to clear the land for coffee production and pollution from pesticides.

How fair trade has helped coffee producers

- Fair trade guarantees a minimum wage for the harvests of small producers; this means they can provide for the basic needs for their families.

- Farmers are provided with credit facilities and paid a minimum price.

- Fair trade brings the product directly to the consumers and cuts out some of the intermediate costs of middlemen so that the benefits of trade are more likely to reach the producers.

- Fair trade develops long-term trade based on trust and respect.

- Fair trade works with co-operatives so that the producers control the business and the members of the cooperative share the profits and benefits fairly.

- The profits from receiving fair wages can be used by the producers to invest in health, education and the protection of the environment.

Coffee producers in Mexico

- In the village, farmers run a bus service to the nearest town which takes their children to the only secondary school.

- The farmers have bought a computer to track the sales of coffee and keep records on the coffee crop.

- Fair trade encourages *sustainable* farming practices, e.g. in Mexico organic farming, which does not use chemicals, and sustainable cultivation methods have been used to grow the coffee.

Advantages of fair trade to MEDCs

- If producers in the LEDCs earn higher wages and are helped to develop in the long term, they will be able to increase their spending power. This means they will be able to afford to buy high-value processed products such as computers from the MEDCs. In turn this means the MEDCs will expand their trade and will have new markets in which to sell their manufactured products. This is becoming more important as 75% of the world's population lives in the LEDCs.

- Consumers in MEDCs will be able to buy top quality products knowing that the trade has been good for everyone.

- Consumers in MEDCs are able to keep their conscience clear as they know they have helped poor producers in LEDCs by trading in a moral way; people will feel they are being good citizens by caring for others. Trading in a fair way will contribute to people's well-being and development in other countries so improving the quality of life for everyone.

activities

Complete the following using Figs 48–51.

1 Describe how the world trading system treats producers in LEDCs. [6]

2 State the meaning of the term fair trade. [2]

3 Explain how fair trade benefits *individual producers* in LEDCs. [5]

4 Describe two ways in which fair trade benefits *communities* in LEDCs. [6]

5 The consumer has a lot of power in the world of trade – if you buy a product it will stay in the shop. Suggest two ways in which *you* could support fair trade. [4]

HIGHER TIER

6 Imagine you are the publicity officer for an organisation campaigning for fair trade. Write one A4 page article for your organisation's magazine. Outline the reasons for your campaign. Use ICT to present your article. [10]

7 Find out how the Co-op and Craigavon Borough Council are helping to promote fair trade. [6]

8 Describe the message the makers of Divine milk chocolate are trying to convey through their design and packaging of their chocolate bars. [5]

AID

Types of aid

Aid is the giving of resources by one country or organisation to another country. Resources can be in the form of:

- Money – loans and grants;
- Expertise – people who have skills and knowledge, for example engineers and teachers;
- Goods – food, technology and equipment.

Aid can also be subdivided into short- and long-term aid.

- **Short-term aid** is also known as emergency aid and is aid given in response to a particular immediate need, for example after natural disasters such as floods and earthquakes.

- **Long-term aid** is aid that usually takes years before it is of benefit to a country, for example improved education or a tree planting scheme.

Why is aid needed by LEDCs?

LEDCs need aid to help them cope with the following:

- Global inequalities due to differences in levels of development;
- Introducing sustainable methods of development;
- The imbalance of trade which causes a trade deficit;
- Dealing with environmental disasters, for example earthquakes, floods and famine;
- Recovering from man-made disasters such as civil war.

Figure 52
Different types of aid.

	Description	Examples	Advantages/Disadvantages
Bilateral Aid	Aid given directly from one country to another. It is sometimes 'tied' with strings attached.	Ireland prioritises aid to least-developed countries, for example Mozambique. This represented 50% of Ireland's bilateral aid in 1998.	■ Large schemes use up land belonging to the local people. ■ Aid can lead to debt as LEDCs often cannot afford to pay back loans. ■ Aid can encourage corruption so often the most needy do not receive the aid. ■ Aid money is often spent on prestigious schemes, for example dams which do not benefit the majority. ■ Can foster strong links between donor and recipient countries.
Tied Aid	With tied aid the donor country specifies what the aid should be spent on, for example health, education, etc. Aid can also be tied to trade patterns, for example the money must be spent on goods from the country giving the aid.	Some countries offer aid to those trading in arms to their advantage or where there are favourable trade patterns.	■ Aid can lead to debt as LEDCs often cannot afford to pay back loans. ■ Tied aid forces the recipient to buy goods – arms, manufactured goods from the donor.
Multilateral Aid	In this, governments donate money to world organisations or agencies and these bodies then distribute the aid.	The World Bank. The European Commission.	■ Although farming and industry are often targeted, products are often sent abroad to MEDCs. ■ LEDCs can become dependent on aid and fall into more debt. ■ Helps LEDCs to develop new forms of agriculture and industry.
Voluntary Organisations	These are funded by the general public to organise aid programmes.	Comic Relief, Save the Children, Oxfam, VSO, Action Aid.	■ Not tied. ■ Encourages low-cost self-help schemes which are more likely to be sustainable. ■ More likely to go to those most in need. ■ Deals with emergencies. ■ Dependent on giving from general public. ■ Income uncertain. ■ Encourages development of local skills and use of local raw materials. ■ LEDCs do not get any further into debt.

Reasons for lending aid

Governments and individuals have different motives for giving aid. It can be a **political** decision, for example Egypt is an important ally for America in a strategically significant position in the Middle East. This friendship was important to the Americans in the 2003 war with Iraq. The relationship between the two countries is strengthened by the fact that America lends Egypt more aid.

Countries also give aid for **economic** reasons, for example some MEDCs may see the potential in maintaining links with LEDCs, many of whom may be ex-colonies, because of the large market they provide. Many projects also require longer term assistance, for example the UK donated £100 million to assist in building the Victoria Dam in Sri Lanka. Many UK scientists and engineers were employed in constructing the dam.

Many individuals and countries feel they have a **moral** obligation to assist those countries facing difficulties or having limited financial reserves of their own.

An example of aid – Comic Relief

Comic Relief is a well-known charity which uses comedy and laughter to support hundreds of groups, organisations and charities across the UK and Africa to tackle poverty and social injustice. It raises money from the public by involving them in events and projects that are innovative and fun. Comic Relief also seeks to inform and educate the public on issues relating to social change, working hard to make sure that the projects it supports are well thought through, well managed and will make a real difference to people on the ground. Comic Relief is committed to help end poverty and social injustice. Since 1985 when it was started, Comic Relief has raised more than £310 million from Red Nose Days and other events including Sport Relief in 2002.

In 2003 the Red Nose Event was The Big Hair Do. Events like this appeal to all ages and are a fun way of raising money for aid projects.

> **Top Tip**
>
> Why not get your school involved? You could organise a fund-raising event and your teacher can order a teaching pack. Check out the web for more details.

Below are two projects recently supported by Comic Relief.

Tanzania

Why they need help?

Since the early 1990s, Tanzania has witnessed a real increase in the number of children living and working on the street. This is largely because of poverty and HIV/AIDS. Not only do boys and girls who live on the streets have difficulty accessing basic services such

Figure 53
Comic Relief 2003.

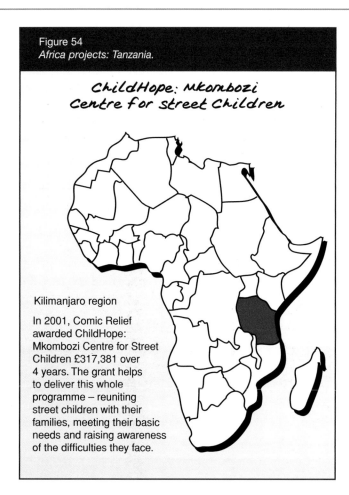

Figure 54
Africa projects: Tanzania.

ChildHope: Mkombozi Centre for Street Children

Kilimanjaro region

In 2001, Comic Relief awarded ChildHope: Mkombozi Centre for Street Children £317,381 over 4 years. The grant helps to deliver this whole programme – reuniting street children with their families, meeting their basic needs and raising awareness of the difficulties they face.

as healthcare and schooling, but also they are often verbally, physically or sexually abused. As a result, many do not trust adults. Most street children have not had any formal education, which makes it hard for them to earn money legally and so many are forced into crime to feed themselves. This can often lead to drug and alcohol misuse and even prostitution.

What is being done

Through their Family Reunification and Community Outreach Programme, Mkombozi aims to tackle the causes of child homelessness and to offer urgent help to children who are living on the streets. Mkombozi meets with street children and find out how and why they have ended up on the streets. They then try and tackle the problems of each child, visiting the family and meeting with community leaders to try and find a way to reunite them. They have also established a support centre where they offer food, shelter, basic schooling and vital counselling and support. Mkombozi also runs drama groups, producing and performing plays that highlight the reasons for child homelessness and the everyday difficulties that street children face. These plays offer the kids something fun and engaging to do, but also help to raise awareness about the plight

of street children within the wider community. Each year Mkombozi reaches out to around 400 boys and girls living on the streets in the Kilimanjaro region.

Kenya

Why they need help

Because of rapid urbanisation in many of Kenya's cities, nearly 8 million people have no choice but to live in poor shelter in overcrowded areas. It is estimated that more than 60% of Nairobi's population live in slum areas on just 5% of the city's land. Most have no access to basic services such as safe water or sanitation. It's estimated that over 90% of Nairobi's slum dwellers do not have access to adequate sanitation and many poor families have to buy water from vendors, which is very expensive. Diseases spread fast and infant mortality rates are incredibly high. Local councils ignore the desperate need to improve the living conditions of these people and they threaten them with eviction. Kenya's slum dwellers are poor and incredibly vulnerable.

What is being done

NACHU is a housing union established in 1979 in response to the need for decent housing for poor

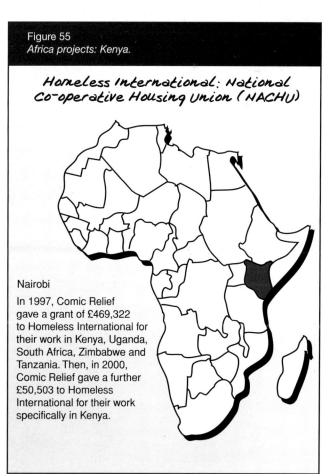

Figure 55
Africa projects: Kenya.

Homeless International: National Co-operative Housing Union (NACHU)

Nairobi

In 1997, Comic Relief gave a grant of £469,322 to Homeless International for their work in Kenya, Uganda, South Africa, Zimbabwe and Tanzania. Then, in 2000, Comic Relief gave a further £50,503 to Homeless International for their work specifically in Kenya.

Kenyans. A Comic Relief grant in 1997 enabled NACHU to improve poor people's basic living conditions by establishing a revolving loan scheme. Loans are given to people living in slums to buy land on which they can build safe homes for the first time in their lives. Once they've paid the loan back, the money is given to another person, so the money is used time and time again. The benefits of these loans reach beyond individuals and families to whole communities. Whilst managing the loan scheme, NACHU offers support, advice and training to make sure the scheme works effectively. A unique feature of the programme has been the 'community exchanges'. This has enabled poor communities to make visits to others in similar situations, to learn from each other's experiences. NACHU will also help people to negotiate with local authorities about their neighbourhood's needs and vital facilities such as water and health.

Positive and negative outcomes of aid

As can be seen above, aid can have positive benefits for recipient countries, promoting development and improvements in standards of living and the quality of life of many, while enabling people to help themselves. However, it can also have negative outcomes. It doesn't always reach the people in most need, thereby widening differences. It can be used for overambitious schemes and some feel it can lead to overdependence on MEDCs.

Figure 56 shows some of the problems associated with aid, which can lead to negative outcomes, or the aid being ineffectual. Care must be taken to ensure that:

- Aid programmes involve local people;
- Aid is appropriate to the situation;
- We do not assume that all types of aid are useful to developing countries.

web link & extra resources

www.comicrelief.com

Figure 56
Problems with aid.

activities

1   Study Fig. 56 'Problems with aid'. Match the cartoons with the labels below:

■ Aid doesn't reach the needy

■ Aid creates debt

■ Problems with the distribution of the aid

■ Problems with the use of the product

■ Aid is unsuitable

■ Aid is tied

■ Export earnings are needed to pay off debts

■ Aid undermines local producers

■ Aid creates dependence

[9]

2   (a) What is the difference between tied and voluntary aid?

(b) Give two examples each of short-term and long-term aid.

(c) What are the benefits of long-term aid to a LEDC?

(d) State fully three disadvantages aid can bring to a LEDC.

[15]

3   (a) Match the problems A–H below with the solutions 1–8 overleaf.

(b) Suppose you had £20 to give to one of the solutions, which project would you select and why?

(c) Read the whole story situations overleaf which correspond to the solutions 1–8 and list the ones which are appropriate types of aid.

(d) Explain why some types of aid are inappropriate.

[20]

Aid problems:

Burkina Faso: A

Soil is being eroded from farmland as a result of fast flowing water.

Liberia: B

As a result of civil war many Liberians fled from their homes and lived in camps. They lacked many things including food.

Ghana: C

Babies that are born prematurely are more likely to die.

India: D

Children's education is suffering as their is a lack of textbooks in schools. There is a great need for secondary level science textbooks.

Guinea Bissau: E

There is a shortage of fresh water for local villagers.

Guinea Bissau: F

Farmers have no way of transporting their produce to market. Public transport is too expensive and to carry it by hand is too difficult and takes too much time.

Nepal: G

There are many landslides which frequently block roads and destroy agricultural land.

Tanzania: H

Coconut oil is an important income provider for many women. The graters used for producing the oil often don't work and are painful to use.

Aid solutions:

A baby incubator *(labelled 1)*

A pump well *(labelled 2)*

Donkey and cart *(labelled 3)*

Sacks of wheat *(labelled 4)*

Building stone banks *(labelled 5)*

Planting trees and plants *(labelled 6)*

Research for a new design of coconut grater *(labelled 7)*

Science textbooks *(labelled 8)*

Aid whole story situations:

1
The baby incubator cannot be used as the hospital's electricity supply is not constant. There also has been no training for African staff in how to use such a complex machine.

2
The pump wells were installed by the Japanese. The wells are made in Japan and therefore use Japanese parts. The pumps have broken down and the locals cannot get new parts to fix the pumps, as they were not involved in installing them.

3
The carts are made by a local tradesman and so can be repaired easily when necessary. The donkey and cart gets the farm produce to market while it is fresh and in good condition. This means a better price for the farmer.

4
The Canadian government sent wheat to Liberia which is a rice eating country. People did not know how to cook with wheat nor did they like it.

5
The women farmers were involved in building the stone banks that have saved water and slapped the soil washing away. The land has been reclaimed and harvests have greatly increased.

6
Local engineers are being trained by overseas engineers. The engineers work alongside local farmers who have the greatest knowledge about the vegetation. As a result the farmers learn how to save the soil and keep the roads clear.

7
Research has provided a new design of grater that is easy to use and increases coconut oil production. The grater is manufactured in a local workshop. The women can afford the new grater as they can pay for it in weekly instalments.

8
The textbooks were sent to a school in India by a school in the UK as they were no longer being used. The books are in English so only well-educated Indians can read them. Also, they are old and the information is badly out of date.

4 Study Fig. 57.

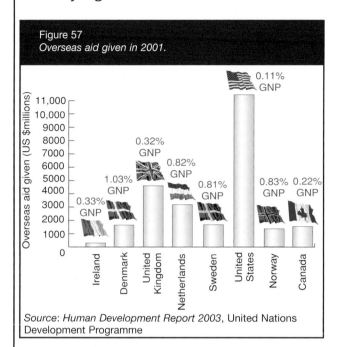

Figure 57
Overseas aid given in 2001.

Source: *Human Development Report 2003*, United Nations Development Programme

(a) Which country gave the most aid in millions of US$ in 2001?

(b) How much money did the USA give as aid to other countries?

(c) Which country gave the largest part of its total income to overseas aid?

(d) Explain how the country which gives the most money is actually being the least generous in its giving to overseas aid.

[6]

5 Study Fig. 58.

Figure 58
Official aid received in 1999.

Top Ten Recipients of Official Aid in 1999			
	HDI – 1999	**US$ m**	**US$ per capital**
China	0.72	2324	1.9
Indonesia	0.68	2206	10.7
Russia	0.78	1816	12.4
Egypt	0.64	1579	25.2
India	0.57	1484	1.5
Vietnam	0.68	1421	18.3
Bangladesh	0.47	1203	9.4
Thailand	0.76	1003	16.7
Tanzania	0.44	990	30.1
Poland	0.83	984	25.5

Source: *Human Development Report 2001*, United Nations Development Programme

Explain why China, India, Vietnam and Egypt may have received so much aid in 1999.

[6]

Top Tip

Remember most of the poorest, least-developed countries do not appear on the table.

WHAT MAKES A GOOD SITE FOR A SETTLEMENT?

The **site** of a settlement is the actual place where a settlement is built. Most settlements have an older centre, which often includes its original site. Many of the settlements in Ireland are thousands of years old, and their sites reflect the needs of people at that time. Most of these were physical factors to do with the shape of the land (relief), closeness to water, shelter and defence.

Three of the most common settlement sites

■ Wet point site

Here closeness to a water supply is the most important factor influencing the **location** of the settlement. Before piped-in water supplies, like today, people had to fetch water by hand and carry it to their homes. This meant that in drier areas it made sense to build houses close to the water, to minimise this labour. In limestone areas of Ireland many settlements were located at the resurgence springs found at the foot of limestone escarpments.

■ Defensive site

Many settlements were established by a single tribe, and sometimes it was important to have your settlement on land that could be easily defended form attacks launched by neighbouring tribes. Good **defensive sites** include the top of hills or the inside a wide meander bend.

■ Bridging point site

Such settlements are on a spot where it is easy to cross a river. For example, Belfast has this kind of site. The site of Belfast is close to the Queen's Bridge, next to the newly built Lagan Lookout. Here a large sandbank was exposed at low tide that allowed travellers to ford the Lagan when this happened. Crossing points of main rivers are worth defending, and when the Normans invaded in the late 12th century they built a castle near the **bridging point**. The castle has long since disappeared, but the surrounding streets still remind us of its existence, e.g. Castle Place.

Today a settlement's original site characteristics can be traced from their names (see table on page 171).

Figure 1
Choosing the best site for a settlement.

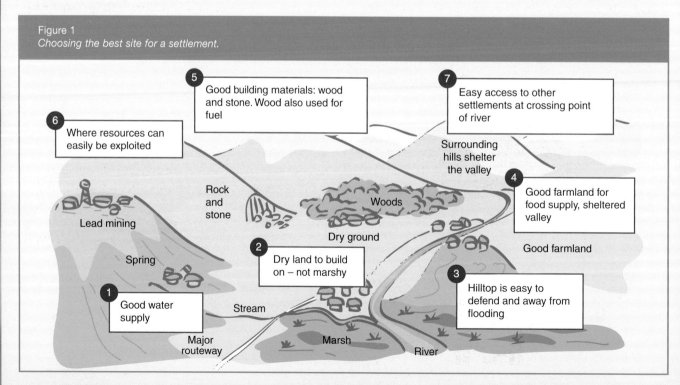

Ard	height or high
Ath	a ford in a river or stream
Beag	small
Dun	a fortress
Daire	an oak grove or wood
Druim	a large ridge or long hill
Gleann	a glen or valley
Cnoc	a hill
Mor	big or great
Sliabh	a mountain

Settlement growth

Once a settlement was established, the rate and direction in which it grew depended on a combination of physical and human factors.

Looking at a map of a settlement, along with photographs, can help chart its growth.

Since settlements grow outwards from a central point, the greatest concentration of oldest buildings in settlements are found in their city centre. Depending on the physical and human features of an area, many small settlements grew into towns or even cities because of the advantages they had over other local settlements. As places grow, these advantages often created the main **function** of the settlement.

Figure 2
Growth of a typical town settlement.

OLD TOWN (BEFORE 1837)

ON A MAP

Some old towns grew up around an abbey or castle

Cathedral

Churches with towers or steeples

Blocked in area of streets: no regular pattern

Try to locate the centre of the town. Clues include pedestrian zones, a town hall or cathedral. In very old settlements there may be buildings that date back to the 1400s. In Belfast there are good examples of Georgian buildings. Newtownards has one of Europe's best preserved Georgian town halls.

VICTORIAN (1837–1901)

ON A MAP

Works

Works

Before the coming of cheap public transport workers had to live within walking distance of the factories and mills

Rectangular grid of streets (bends not necessary before motor vehicles)

Railway station on fringe of town in 1850s, or whenever it was built, stimulated growth of town close to station

The Victorians were prolific builders. Their legacy can be seen in all of the UK industrial cities. Belfast was very important during this period, and many of the inner city and inner suburbs were built during the late 1800s. Look out for red-brick buildings.

20th + Early 21st Century (1901 to present day)

ON A MAP

Cul-de-sacs

Housing estates often filling in gaps between main roads

Less densely built up than in Victorian streets

Carefully planned

You should be familiar with modern-day developments. Houses in the outer suburbs have such street plans. They are designed to reduce the amount of through traffic, so that children can play safely. These plans take up a lot of space, so are more often seen on the outer suburbs, where land is cheaper.

What types of function can a settlement have?

- Market centre – where goods are bought and sold. Originally this would have been at a weekly market, although now it is mostly retail through permanent shops.

- Port – where ships use the port facilities of the settlement for importing and exporting goods.

- Industrial centres – settlements with many factories that make goods.

- Route centres – these develop where several roads meet or at nodal points created by economic development.

- Tourist centres – where economic activity in the settlement is strongly linked to tourism. They may be mountain resorts, coastal or even spas.

- Cultural centres – attract people for educational or religious reasons.

- Mining towns – a local resource that can be mined attracts workers and wealth, allowing the settlement to flourish.

- Dormitory town – where most residents work elsewhere.

Cities can grow and engulf (swallow-up) several surrounding towns, creating a large urban area known as a conurbation. This name comes from the phrase – continuous urban area. If a city grows so much it meets another city, they create the biggest type of urban area, known as a megalopolis. One example of this is on the eastern coast of the USA where Boston, New York and Washington are linked together to form a megalopolis nicknamed 'Boswash'. This area contains over 40 million people and the lights from this megalopolis can be seen from the moon.

In order to meet demand for housing and services, the city is always eating into the surrounding countryside. Planners and the government agree that sprawling cities are unsustainable, so green belts are imposed. These are zones around a city that have very strict planning restrictions so that building is controlled. Belfast has a green belt area, bounded by a line called the Mathew Stop Line. To preserve the rural–urban fringe, the latest planning ideas involve the regeneration of the city's inner-city area. Current recommendations from the Belfast Urban Area Plan are that over 50% of new houses should be built within the existing Belfast urban area.

activities

1   State fully the meaning of the term 'site'. [2]

2   Create a learning map, to show the factors that influence the site of a settlement. [7]

3   Study the diagram below which shows four settlements in Northern Ireland. Answer the question that follows.

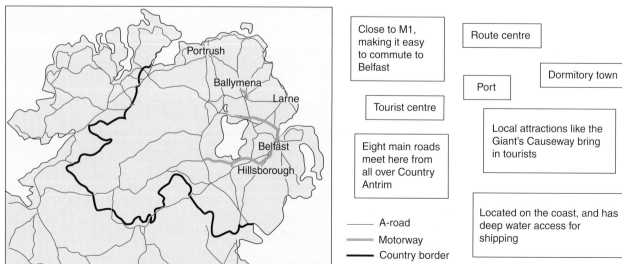

Using an atlas map of Northern Ireland to help you, complete the following table about the four settlements using the boxed information above.

Name of settlement	Situation characteristics	Main function of settlement
Ballymena		
Hillsborough		
Larne		
Portrush		

[8]

HIGHER TIER

4 (a) Using an atlas map of Northern Ireland to help you, add in examples to the table of the other four types of function. [8]

(b) Study a map of Belfast and the diagram below. Identify and explain the main locational factors related to its site. [5]

Site of Belfast

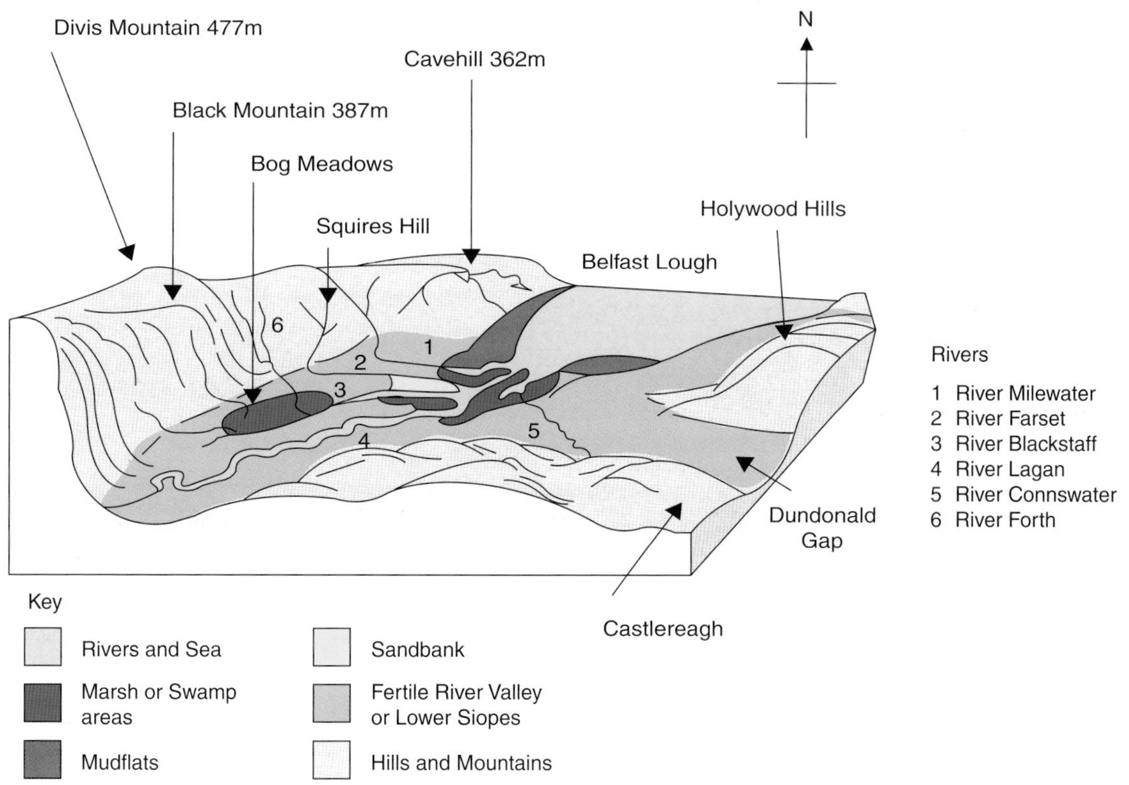

Divis Mountain 477m

Cavehill 362m

N

Black Mountain 387m

Bog Meadows

Squires Hill

Holywood Hills

Belfast Lough

6

1

2

3

5

4

Rivers

1 River Milewater
2 River Farset
3 River Blackstaff
4 River Lagan
5 River Connswater
6 River Forth

Dundonald Gap

Castlereagh

Key

Rivers and Sea

Sandbank

Marsh or Swamp areas

Fertile River Valley or Lower Slopes

Mudflats

Hills and Mountains

SETTLEMENT HIERARCHY

A settlement **hierarchy** is when settlements are placed in order of importance. The most important is placed at the top of the hierarchy. The importance of a settlement is determined by examining its size, sphere of influence, and the number and type of services it provides. Generally speaking, the largest settlement with the most services is seen as more important than the smallest settlements that offer few or no services at all.

(The size of a settlement can be determined by its area or the number of people living in it.)

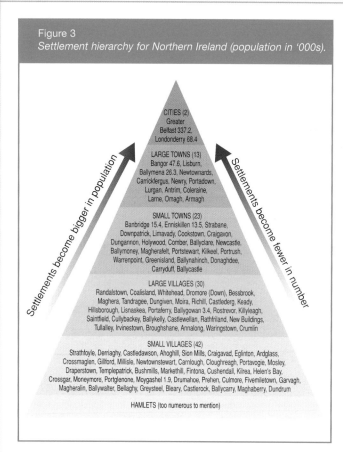

Figure 3
Settlement hierarchy for Northern Ireland (population in '000s).

CITIES (2)
Greater
Belfast 337.2,
Londonderry 68.4

LARGE TOWNS (13)
Bangor 47.6, Lisburn,
Ballymena 26.3, Newtownards,
Carrickfergus, Newry, Portadown,
Lurgan, Antrim, Coleraine,
Larne, Omagh, Armagh

SMALL TOWNS (23)
Banbridge 15.4, Enniskillen 13.5, Strabane,
Downpatrick, Limavady, Cookstown, Craigavon,
Dungannon, Holywood, Comber, Ballyclare, Newcastle,
Ballymoney, Magherafelt, Portstewart, Kilkeel, Portrush,
Warrenpoint, Greenisland, Ballynahinch, Donaghdee,
Carryduff, Ballycastle

LARGE VILLAGES (30)
Randalstown, Coalisland, Whitehead, Dromore (Down), Bessbrook,
Maghera, Tandragee, Dungiven, Moira, Richill, Castlederg, Keady,
Hillsborough, Lisnaskea, Portaferry, Ballygowan 3.4, Rostrevor, Killyleagh,
Saintfield, Cullybackey, Ballykelly, Castlewellan, Rathfriland, New Buildings,
Tullalley, Irvinestown, Broughshane, Annalong, Waringstown, Crumlin

SMALL VILLAGES (42)
Strathfoyle, Derriaghy, Castledawson, Ahoghill, Sion Mills, Craigavad, Eglinton, Ardglass,
Crossmaglen, Gillford, Milisle, Newtownstewart, Carnlough, Cloughreagh, Portavogie, Mosley,
Draperstown, Templepatrick, Bushmills, Markethill, Fintona, Cushendall, Kilrea, Helen's Bay,
Crossgar, Moneymore, Portglenone, Moygashel 1.9, Drumahoe, Prehen, Culmore, Fivemiletown, Garvagh,
Magheralin, Ballywalter, Bellaghy, Greysteel, Bleary, Castlerock, Ballycarry, Maghaberry, Dundrum

HAMLETS (too numerous to mention)

Settlements become bigger in population

Settlements become fewer in number

ORDERS OF GOODS AND SERVICES

It is useful to put goods and services into one of three categories by looking at how often we use the service.

Low order – these are goods and services that we buy or use daily or every other day, e.g. buying milk or going to post a letter in a post box.

Middle order – are goods and services that we would buy or use on a monthly or fortnightly basis, e.g. buying clothes or going to the cinema.

High order – these are goods and services that people only buy or use rarely, maybe once a year, or even less, e.g. buying a new sofa or going into hospital.

We can also examine goods and services by looking at how far people will travel to use the service and how many customers are needed to keep the service profitable.

The **range** of a product or service is the maximum distance people will travel to buy or use it. Convenience goods have a short range. Comparison goods have a longer range.

The **threshold** of a product or service is the minimum number of sales needed to make a profit. It is possible to state a threshold value for each good or service, e.g. a small grocers shop needs 350 customers, a secondary school 10,000 local residents and Sainsbury's 60,000 customers.

Settlements higher up the hierarchy provide more services than those lower down on the hierarchy. Settlements at the top of the hierarchy also provide more high-order services as well as low- and middle-order services. In relation to retail, high-order services often include comparison goods. Low-order services may include convenience goods. As we move up the settlement hierarchy, each level provides all the services offered at the level below, as well as new services.

Figure 4
Settlement hierarchy by service.

Capital city or conurbation	Cathedrals, large hospital, national sports stadium and teams, out-of-town superstores, international airports, national museums and galleries, main local government buildings and all other services listed below.
City	One cathedral, full range of shops including department stores, main bus and train stations, large hospitals, university, local airport, court buildings, theatres, national sports teams and all other services listed below.
Large town	Wide range of shops, including the well-known high street shops, shopping centres, supermarkets, local bus and/or train station, small hospital, hotels, banks, opticians, furniture and electrical stores, solicitors, main police station and all the services listed below.
Small town	Small supermarket, health centre, dentists, several churches/chapels, restaurants, bus station, bank, small secondary school and all the services listed below.
Village	One or two churches/chapels, public house, primary school, convenience goods shop, bus stop, small post office, village hall and all other services listed below.
Hamlet	Bus stop, post box and public telephone, if any services at all.

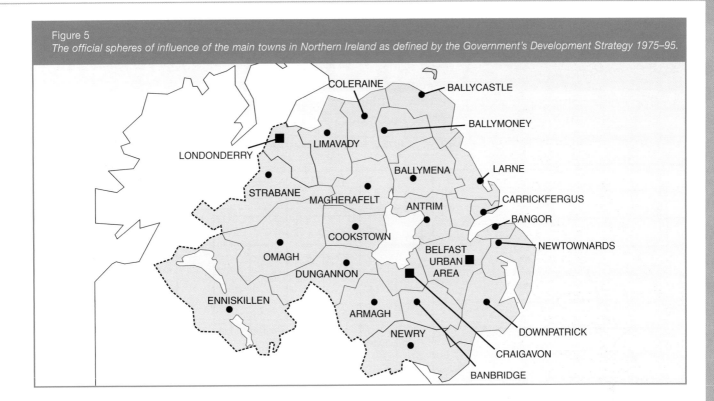

Figure 5
The official spheres of influence of the main towns in Northern Ireland as defined by the Government's Development Strategy 1975–95.

Each settlement attracts people from the land around them to use the services offered in the settlement. This area is known as the **sphere of influence** of the settlement. Large settlements that offer a wide variety of high-order goods and services attract people from a wide area, because the closer, smaller settlements do not offer the high-order services or variety that people want. Thus, large settlements often have a large sphere of influence. Other factors can affect the size of the sphere of influence that a settlement. These include:

The function of a settlement

If tourism is a main function of a settlement, it can draw people from a larger than expected area, as people travel long distances to see the attractions of the settlement. Alternatively, if a settlement is a dormitory settlement, it may provide fewer than expected services for its size, and have a smaller than expected sphere of influence.

Accessibility

If a settlement is easy to travel to, in other words, it is accessible, then it may have an enlarged sphere of influence. This could relate to road and rail connections or even the number of bus routes it has.

Level of competition from rival settlements of the same size or larger

If a town is close to a city, people will use the city as it has the best variety and number of high-order

goods, therefore reducing the size of the town's sphere of influence.

Ways of assessing the sphere of influence of a settlement

- Find out where people have come from to shop in the town centre.
- Look at the catchment area of the largest school in the settlement.
- Examine bus routes.
- Find out the boundary of the authority of the local council.
- Look at the sources of adverts placed in the local newspaper.
- Find out who buys the local paper.
- Find out the service areas of the police station, hospital and fire station.

activities

1   Copy and complete the diagram of a settlement hierarchy shown below. Get your answers from the list of words/phrases below:

Large   Few or none   All orders   Close together   Far apart   Small   Only low order

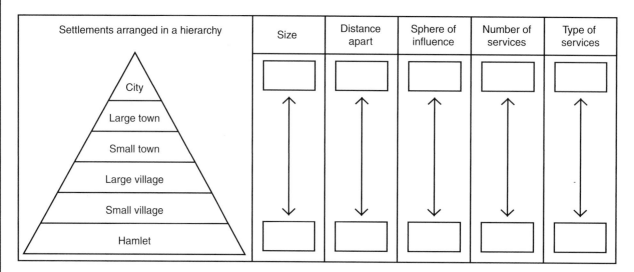

[10]

2   Using data from Fig. 3, use a spreadsheet package to create a bar chart to illustrate the average population sizes of each settlement level in the hierarchy.   [10]

3   Classify the following goods and services into high, middle and low order:

MILK   TELEVISION SET   BANK   NEWSPAPER   JEWELLERS   CLOTHES   POST OFFICE

HIGHER TIER

4   (a)   State three services that might be found in a village.

(b)   State three services that might be found in a city, but not in a village.

(c)   Explain why one of services you gave as an answer for part (b) would not be located in a village.

(d)   Study Fig. 5, which shows the spheres of influence of Northern Ireland settlements. Answer these questions:

(i)   Three of the following statements are true. Pick them out and copy them down.

■ Belfast has a larger sphere of influence than Newtownards.

■ Towns in the west of Northern Ireland tend to have larger spheres of influence than those in the east.

■ All towns have exactly the same sized sphere of influence.

■ Towns close to Belfast seem to have smaller spheres of influence.

■ Enniskillen has a smaller than expected sphere of influence for a town.

(ii)   Choose one of the true statements and explain why that situation exists.

[15]

The distribution of settlements can be affected by physical and human factors

If landscapes were completely flat without any interrupting features, such as roads, rivers and coastlines, then settlements would distribute evenly.

This would create a hierarchy of spheres of influence and all settlements of the same order would be similar distances apart. In reality, features such as mountains, rivers and roads distort this regular pattern, and local settlement **distribution** patterns are established.

Look at these contrasting examples from Co. Antrim.

EXTRACT A. OS Map 1:50,000.

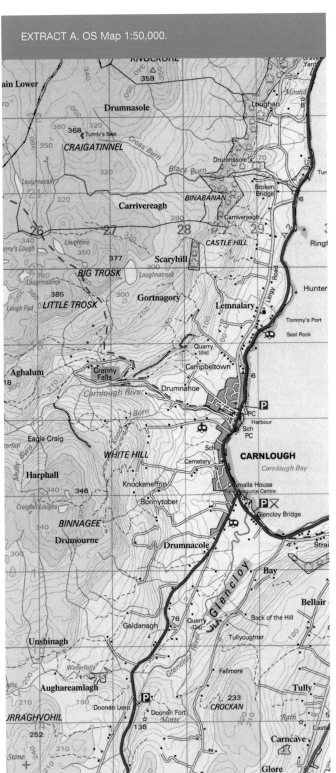

EXTRACT B. OS Map 1:50,000.

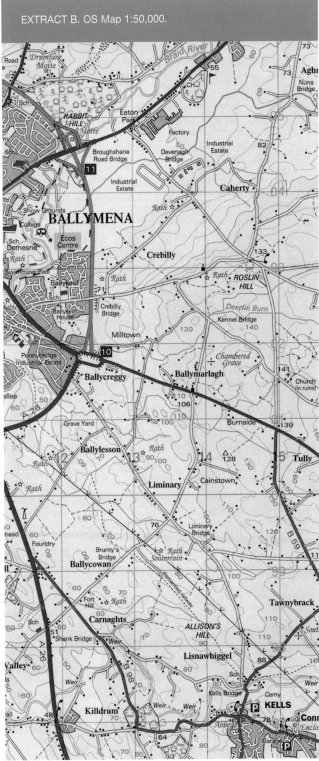

activities

1 (a) State what is found at the following six-figure grid references:

   (i) 297 222

   (ii) 291 194

   (iii) 280 189 [3]

   (b) Give the six-figure grid references for the three caravan parks shown. [6]

   (c) How far is it from the top of Little Trosk to the top of Big Trosk? [2]

   (d) If you were to walk from the farm at 272 179 along the track that follows Knock Burn, would you be going uphill or downhill? State fully how you reasoned your answer.

   (e) State the name of the minor road that goes north-north-east out of Carnlough. [1]

2 (a) Using some of the words from the box below, complete the following paragraph describing the settlement distribution shown on extract A.

   roads evenly 2 km largest 3 km farms coastline smallest cities

   In Extract A, settlements are not _____ distributed over the landscape. Instead they follow the _____ in a linear pattern. Carnlough is the _____ settlement shown, and is a linear village on the coast, stretching just over _____ from north to south, but is only 0.3 km at its widest point east to west. Other isolated _____ follow the river valleys and _____ which go north-eastwards to the coast. [3]

3 (a) Complete the tally chart below that shows the services offered by Carnlough.

Service	Tally
Caravan site	
School	
Church	
Car park	
Post Office	
Picnic site	
Public toilet	
Outdoor centre	

   (b) Using evidence from your tally chart, place Carnlough on the settlement hierarchy and explain your choice. [4]

HIGHER TIER

4 State fully how **three** of the following factors have influenced settlement distribution on the map extracts shown:

   coastlines, mountain area, river valleys, roads. [9]

web link & extra resources

www.irelandstory.com/map_index.html – a great resource of maps showing settlement distribution in Ireland today and in the past.

UNIT TWO
Urban Growth and Change

URBANISATION

Urbanisation is the process by which an increasing percentage of people are living in towns and cities.

For the coming decades, the predicted increase is 80 million new city dwellers every year. The majority of those people will be living in cities that are developing countries. The 21st century is the first in history with the majority of the worlds inhabitants living in urban environments.

The urban population of the world is not spread evenly amongst all cities. Cities with a million or

Figure 6
The world's 25 largest cities, listed in their order of size in 1990 and their positions on the globe.

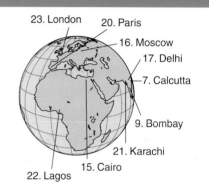

City	Country	1950	1990	2000
1. Mexico City	Mexico	3.1	20.2	25.6
2. Tokyo	Japan	6.7	18.1	19.1
3. São Paulo	Brazil	2.4	17.4	22.1
4. New York	USA	12.3	16.2	16.8
5. Shanghai	China	5.3	13.4	17.0
6. Los Angeles	USA	4.0	11.8	15.7
7. Calcutta	India	4.4	11.8	15.7
8. Buenos Aires	Argentina	5.0	11.5	12.9
9. Bombay	India	2.9	11.2	15.4
10. Seoul	South Korea	1.0	11.0	12.7
11. Beijing	China	3.9	10.8	14.0
12. Rio de Janeiro	Brazil	2.9	10.7	12.5
13. Tianjin	China	2.4	9.4	12.7
14. Jakarta	Indonesia	2.0	9.3	13.7
15. Cairo	Egypt	2.4	9.1	1.8
16. Moscow	Russia	4.8	8.8	9.0
17. Delhi	India	1.4	8.8	13.2
18. Manila	Philippines	1.5	8.5	11.8
19. Osaka	Japan	3.8	8.5	8.6
20. Paris	France	8.0	8.5	8.6
21. Karachi	Pakistan	1.0	7.7	11.7
22. Lagos	Nigeria	0.3	7.7	12.9
23. London	UK	8.7	7.4	6.6
24. Bangkok	Thailand	1.4	7.2	10.3
25. Chicago	USA	4.9	7.0	7.3

☐ LEDC

☐ MEDC

(Figures are in population millions)

Trends to note:

The biggest cities are getting bigger – the top city in 2000 has over 25 times more residents than the top city in 1800. More and more of the world's largest cities are to be found in LEDCs – notably Latin America, coastal Africa and South-east Asia.

more residents are called million cities. Some these have grown so much they are becoming mega-cities – with more than 10 million inhabitants.

Where are the world's biggest cities?

Figure 6 lists the world's biggest cities.

Reasons for today's distribution of millionaire cities

LEDCs are the areas now experiencing their industrial boom time. Multinational and transnational corporations are being attracted to these areas by low wage levels, pools of skilled labour, lack of union laws, and favourable government trade and tax regulations. The jobs are pulling people into the cities, so they grow into millionaire cities.

The **urbanisation** trend is slowing in MEDCs. Transport is easy and with cities being perceived as polluted areas with high crime rates, people are choosing to leave cities rather than move into them. The rising cost of land in MEDC cities has also meant that families who want larger accommodation with a garden cannot afford the type of housing they want in the confines of the city, so they are moving into the near rural areas surrounding the city.

A high natural increase in LEDCs means that the total number of people in cities here is also increasing more rapidly than that of MEDC cities, where birth rates are often so low that the people are not replacing themselves by having two children.

activities

1 What makes the 21st century different from all previous centuries regarding the amount of urban dwellers? [1]

2 (a) State the meaning of the term urbanisation. [2]

 (b) How is a mega-city different from a millionaire city? [2]

 (c) State two reasons to account for the growing number of large cities in LEDCs. [4]

3 (a) Using the information in the table below about the world's largest cities, copy and complete the composite bar graphs below. [4]

Year	No. of cities in the top 10	
1900	0	10
1950	3	7
2000	8	2

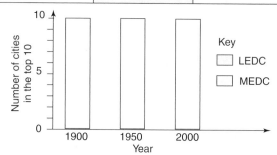

 (b) Describe how the distribution of large cities has changed over the last 200 years. [4]

HIGHER TIER

4 State fully three reasons to account for the changing distribution of large cities across the world. [9]

5 Draw a cartoon to illustrate an aspect of urbanisation, like the one below:

"We're waiting for the city to come to us ..."

Causes of urbanisation

There are two main causes of urbanisation:

- In-migration to urban areas;
- Natural increase of the population in urban areas.

In-migration to urban areas

The growth of Belfast was mainly due to migration. People from the countryside moved into Belfast in the late 19th and early 20th centuries. People's decision to move was based on the prevailing push and pull factors of the day.

Today the population of Belfast is declining due to **counter urbanisation**.

LEDC cities are now experiencing their period of growth, also mainly due to in-migration as people leave the countryside and head to the main city of the area looking for work and a higher standard of living. These migrants are also responding to push and pull factors.

Push factors

- Lack of job opportunities due to mechanisation of agriculture.
- Natural disasters such as drought in Ethiopia or a volcano like Pinatubo erupting.
- Traditional land inheritance system that often means a father must share his farm equally among his sons, so each generation inherits smaller and smaller plots of farmland.
- Lack of government investment in rural areas.
- Harshness of farm life and poor working conditions.
- Problems of crop failure that lead to local famine.

Pull factors

- Better-paid jobs. Factory jobs can pay up to three times what a farmer can earn.
- Better health care in cities.
- Better quality of life due to higher levels of public services such as piped water and gas in cities.
- Increased opportunity for education.
- Hearing stories of success from other rural–urban migrants.

However, with decreasing urban wages and the effects of overpopulation in urban areas, migrants may arrive only to be met with unemployment, hard labour or poor communal living in suburban shanty towns. It may take weeks of tamaking (a Kenyan term that means looking for work) before any promise of employment is found and if one has no relatives or friends in town, living alone or even sleeping on the street may be the only initial options. The expansion of urban areas resulting from this migration has also led to a decrease in the quality of life for many city residents.

This migration, however, only accounts for about 42% of the rise in urban population in LEDCs. The remaining 58% comes from the natural increase of their urban populations, because the birth rate is higher than the death rate. The reasons why birth rates are so high in LEDCs can be found in the chapter on population.

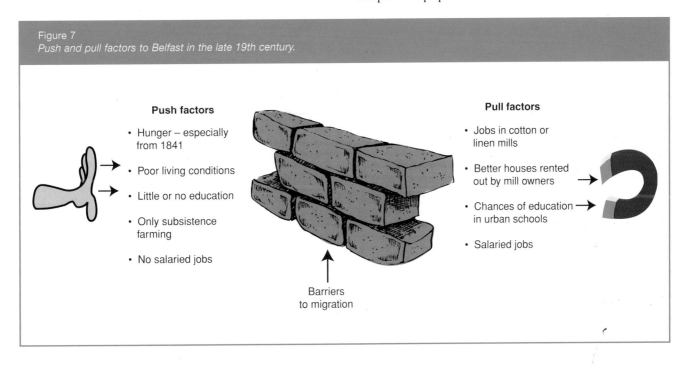

Figure 7
Push and pull factors to Belfast in the late 19th century.

Push factors
- Hunger – especially from 1841
- Poor living conditions
- Little or no education
- Only subsistence farming
- No salaried jobs

Barriers to migration

Pull factors
- Jobs in cotton or linen mills
- Better houses rented out by mill owners
- Chances of education in urban schools
- Salaried jobs

Natural increase

The majority of the population in LEDC cities is made up of younger people. The better medical care available in cities means that babies are likely to survive, meaning that birth rates are considerably higher than death rates, leading to a high natural increase. Thus, cities in LEDCs will continue to grow, even if the rural–urban migration flow stops.

Some general conclusions regarding urbanisation around the world:

- The slower urbanisation is, the easier it is to deal with.
- Rapid urbanisation means rapid increases in the numbers of urban people who need land, housing, water, electricity, health care, and schooling.
- Urban conflicts will be greatest where urbanisation is greatest.
- If the problems of urbanisation are not solved social unrest and environmental pollution will get worse.

Some cities in LEDCs are massive, with huge populations. For the most part this is because the majority of the country's wealth and manufacturing industry concentrated in only one or two of its cities. Internal migration from the countryside to urban areas focuses on these cities, creating giant mega-cities with over 10 million inhabitants. To put this in perspective, remember that Northern Ireland only contains 1.6 million people altogether!

The mega-city often dwarfs other settlements in the LEDC, creating a primate pattern within the hierarchy. This means there is one settlement that is more than twice the size of the next largest city and holds more than its fair share of the country's wealth and power. Mexico City, for example, has over 21 million people, whilst the second largest city in Mexico has just over 2 million.

Current rates of urbanisation are very different. In MEDCs, rather than moving to large urban areas, people are leaving the large conurbations and moving to smaller towns or rural areas. This process is called counter urbanisation.

Figure 8
Growth of two LEDC mega-cities, 1920–2001 (figures in millions).

Year	1920	1940	1950	1960	1970	1980	1991	2001
Rio de Janeiro	1.2	1.8	2.4	3.3	4.3	5.1	5.3	10.1
São Paulo	0.6	1.3	2.2	3.8	5.9	8.5	9.5	24.6

web link & extra resources

www.soton.ac.uk/~peters/er/ied194.PDF – a good essay for Higher Level students on causes of mega-city growth in LEDCs.

www.geography.learnontheinternet.co.uk – a valuable website. Search here for excellent information on urbanisation and counter urbanisation.

www.geographyinaction.co.uk – this has other good information on Belfast

Figure 9
Differences in urbanisation between MEDCs and LEDCs.

Factor	MEDCs	LEDCs
Time	19th century/early 20th century	Post 1950
Industry	Urbanisation linked to the growth of industry during the industrial revolution.	Urbanisation not necessarily linked to industrial change.
Technology	Labour-intensive machinery introduced in new factories at this time.	Technology not necessarily labour intensive – especially if inappropriate.
Employment	Rural migrants moved to city and found work in new factories.	Very few migrants find jobs in the 'formal sector'. Many find jobs in the 'informal sector', e.g. selling fruit on the streets.
Population growth	Death rates fell slowly, then birth rates rose in cities.	Death rates are falling rapidly, but birth rates remain high, resulting in high natural increase in urban areas.
Trade	Many European countries had access to colonies to allow trade and the means to feed the expanding population.	Debt and unfavourable trading relationships. Imported food is expensive.

activities

1 (a) State two push factors and two pull factors to account for people moving from the countryside into the city. [4]

   (b) Study Fig. 7 relating to migration to Belfast. Suggest two possible barriers to migration from the rural areas. [4]

2 (a) Using the information on the table showing the growth of São Paulo (Fig. 8), complete the following sentences:

       _____ people lived in São Paulo in 1920

       _____ people lived in São Paulo in 2001.

       This means that São Paulo has _____ times more people in it today than it did 80 years ago. [3]

   (b) Choose a suitable scale, and draw proportional circles to illustrate the numbers of people living in São Paulo over time. [8]

   (c) Suggest what problems a LEDC city like São Paulo might have in coping with such rapid population growth. Include ideas on housing, school places, medical care and transport. You might want to use the internet to investigate these problems further. [6]

HIGHER TIER

3 (a) State fully three ways the urbanisation process differs between LEDCs and MEDCs. [9]

   (b) Belfast is experiencing counter urbanisation. Visit the census website at www.nisra.gov.uk/census/start.html and write a paragraph outlining evidence to support this statement.

   (c) Explain why the process of counter urbanisation is happening in MEDCs using the following headings:

       Pollution Transportation Crime House prices [8]

LOCATIONS OF LAND-USE ZONES IN CITIES

Settlements have **functional** land-use **zones** and **socio-economic areas** (including **ethnic areas**).

The types of functional land use in a city include commercial/retail, offices, industry, open space/parks/recreational sports areas and residential. Socio-economic areas are defined by the jobs and money earned by the residents of an area. This is reflected in several indicators such as value of the housing, size of housing and number of cars per household. This information is recorded for areas during a census.

Ethnic areas are defined by the people there rather than what the land is used for. They do, however, tend to be residential areas. They may be a concentration of people who earn similar amounts of money (socio-economic), or who are of the same race or practise the same religion.

Urban land-use models

The Burgess model – concentric ring model

Burgess based his model on Chicago in 1924. He claimed that the focal point of a town was the CBD (central business district). As towns developed they grew outwards from the CBD. This means that

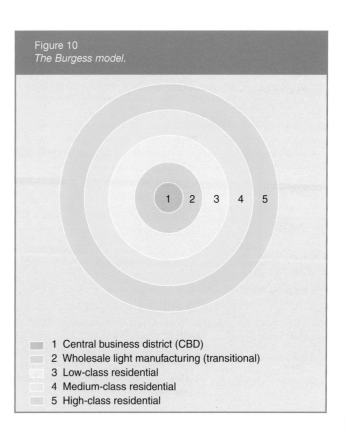

Figure 10
The Burgess model.

1 Central business district (CBD)
2 Wholesale light manufacturing (transitional)
3 Low-class residential
4 Medium-class residential
5 High-class residential

buildings get newer as they move outwards towards the city boundaries. The three housing zones were based on the age of the houses and the wealth of the inhabitants.

Characteristics:

- CBD

  The central business district contains all the major shops and offices thereby making it the centre for commerce and entertainment. All roads lead to it, so it is the most accessible part of the city, therefore many land uses would benefit from being here. As the area it covers is small, and demand for land here is high, **bid rents** are at a maximum in this part of a town or city. This forces buildings upwards and in many CBDs we see skyscrapers.

- Transitional zone/wholesale light manufacturing

  Either invaded by light industry or degrading old housing, containing mostly recent immigrants, so it is a zone of transition; in other words, it is an area that is changing and being redeveloped.

- Low-cost housing

  This is where people who are slightly better off live, these may also be second generation immigrants working in industry. Travelling costs are reduced from this zone, as people can walk to the CBD. It is well served by public transport.

- Medium-cost housing

  Higher quality housing, in UK would include inter-war semi-detached houses and council estates.

- High-cost housing

  Occupied by people who have both the money to afford housing here and who can afford to commute every day.

Problems with the Burgess model:

As it assumes that land is flat it does not show the importance that good roads may have in encouraging settlement. In MEDC cities we are seeing commerce leaving the CBD and inner city to move to cheaper locations, which are now equally accessible as the city centre because most CBDs now have congestion problems. It is also a

model from the 1920s and since then things have changed, in a city such as São Paulo we may find that there are favelas on the outskirts of cities where the poorest people live, who have moved from rural settlements. Since it is based on only one city it only shows that city at one point in time, therefore it would be better to develop a more complex model.

Sector model by Hoyt

This model created in 1939, was based on 142 American cities using eight housing variables for mapping.

Characteristics:

- Wealthy people chose to live where they could afford to, e.g. near services.

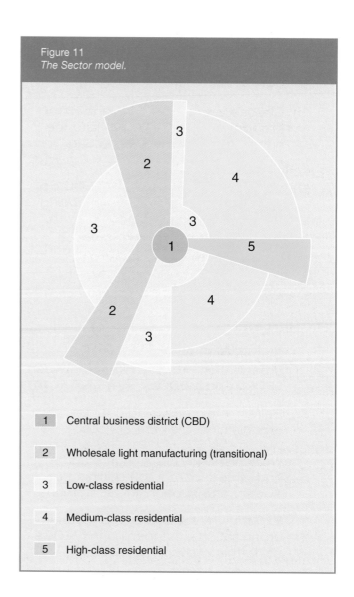

Figure 11
The Sector model.

1	Central business district (CBD)
2	Wholesale light manufacturing (transitional)
3	Low-class residential
4	Medium-class residential
5	High-class residential

activities

1 (a) Copy the table below, and add in three examples of each land use.

Land use	Examples
Retail	
Professional and commercial	
Industry	
Parks/open space	
Residential	
Public buildings	
Other	

[9]

(b) To investigate land use properly, the residential category needs sub-divisions. Suggest suitable sub-divisions for this category. [5]

(c) State fully why 'Other' has been included as a category on the table. [3]

2 State what is meant by the term ethnic area. [3]

3 (a) Make a labelled copy of the two urban land-use models (Figs 10 and 11). [9]

(b) Write three sentences to describe each model. [6]

HIGHER TIER

4 Describe and explain the location of high-class housing and industry in the concentric and sector models of urban land use. [10]

■ Wealthy residents used their cars as transport from home to work and vice versa thereby living further from industry but close to main roads.

■ Similar types of land use clustered together to create 'sector' development, often following road or rail links.

Problems with the model:

Owing to its age this model is also incorrect as it does not take into account commuter villages and changes in the rural–urban fringe which developed with the popularisation of the car. Other problems are similar to those stated for the Burgess model.

Why different functional zones develop

The location of different land-use (functional) zones in a city is related to three main factors:

■ Accessibility;

■ The sequence of urban development;

■ Cost of land.

Accessibility

Commercial land uses cluster in the CBD as this is the easiest place in the city to reach. This is because most road and rail links meet here, so it is equally accessible to people from all over the city. Places on the edge of the city are only easily accessible to local people.

Sequence of urban development

When many European cities and towns began to expand in the 19th century due to rural to urban migration, the main demand for land came from industry, as the industrial revolution was in full swing. To house the workers for the factories a lot of low-cost housing was also required. This was generally provided in the form of terraced housing in Belfast and most other cities in the British Isles.

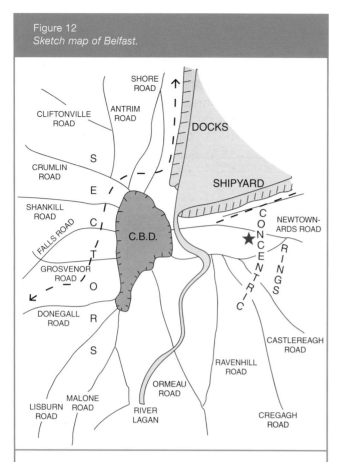

Figure 12
Sketch map of Belfast.

Look at the sketch map of Belfast. There are 16 main roads that lead into the city centre, making it very accessible to people from all over the province.

The star shows the location of the Connswater Shopping Centre in the inner-city area of east Belfast. It is more accessible to local residents in the east of Belfast than to people who live elsewhere. This reduces its potential number of customers.

These new land uses were located next to the old city, creating the zones we see in the models, either rings or sectors of light industry and low-cost housing. Current demand is for high-quality larger housing, retail areas with parking space and industry on landscaped sites. These are located on the edge of the city on what are known as greenfield sites. Land near the city centre has been used up and only in the outskirts of a city is there open space for large building developments. This sequence of development has led to a recognisable city skyline that reflects the age of the land use.

The traditional sequence shown in Fig. 13 is not as applicable as it used to be in the 1970s because in the last few decades there has been substantial urban regeneration that means older housing has been knocked down and replaced.

Cost of land

Land values are not the same across a city. Land in the CBD is the most expensive anywhere in the city. This is because it is the most accessible part of the urban area and so several different types of land use compete to locate there. Because the CBD is also a small area, this high demand pushes up the price of the land. In general, land values decrease as you move away from the CBD. The term bid-rent states the amount of money a land use is willing to pay to occupy a piece of land. This concept of land values falling as distance from the city centre increases is known as

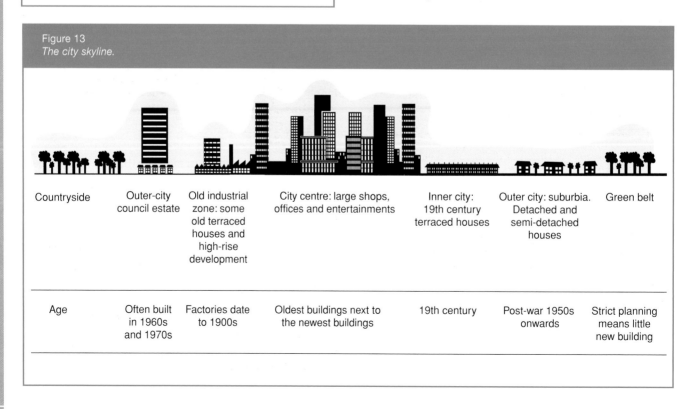

Figure 13
The city skyline.

Countryside	Outer-city council estate	Old industrial zone: some old terraced houses and high-rise development	City centre: large shops, offices and entertainments	Inner city: 19th century terraced houses	Outer city: suburbia. Detached and semi-detached houses	Green belt
Age	Often built in 1960s and 1970s	Factories date to 1900s	Oldest buildings next to the newest buildings	19th century	Post-war 1950s onwards	Strict planning means little new building

distance-decay, and is illustrated on the bid-rent diagram below:

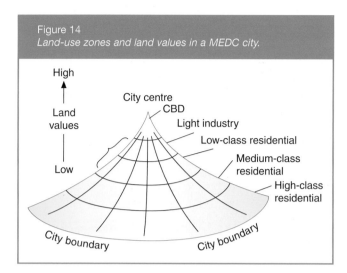

Figure 14
Land-use zones and land values in a MEDC city.

Sprawling cities – the need for green belts

Cities are surrounded by smaller towns, villages and countryside areas, all of which feel the influence of a growing city.

In many MEDCs planners are trying to protect the countryside by imposing green belts around cities. A green belt is an area of countryside that surrounds built up areas and has strict planning regulations against building.

The first green belt was imposed around London in 1944, a city that is still pressurising the countryside around it.

Purposes of green belts:

■ To control the spread of the suburbs;

■ To prevent neighbouring settlements merging;

■ To preserve the character of towns;

■ To protect the countryside against urban sprawl;

■ To encourage urban renewal in the inner city.

Life on the edge of the city – the rural–urban fringe

Just beyond the suburbs, in areas where building is not as strictly controlled as in green belts, some urban development occurs, and the line between city and countryside, urban and rural is blurred. This is the rural–urban fringe.

activities

1   Copy and complete the table below about the factors affecting land-use development in a city. Use the text above and the information from the speech bubbles below the table to help you.

Factor	Meaning of factor	Example from speech bubbles
Accessibility		
Sequence of urban development		
Cost of land		

We want to move into a house with character, built by the Victorians, so we're moving into the inner city

I want lots of main roads near my new shop, so I can attract customers from all over the country

I am going to locate my large factory on the edge of the city, because the land rent will be low there

[6]

2   (a)   State two reasons why large, expensive, modern housing is often found on the edge of cities.   [4]

    (b)   Explain why a golf course is not likely to be found in the CBD of a city.   [3]

HIGHER TIER

3   Research a city or town near you and complete a diagram, similar to Fig. 13, on this settlement. Explain the location of the land uses you include.   [9]

Figure 15
The rural–urban fringe.

Service, retail and warehouse developments

Development of an out-of-town shopping centre. The car park and buildings use up a huge amount of space. Notice the waste land between this and other developments.

Office development

Remains of fields. It is no longer profitable to use these fields because they are becoming surrounded by developments. This is, at present, waste ground.

Single-family housing estate

Condominiums (groups of privately owned apartments). This development includes tennis courts.

Remains of woodland

Road laid out for future development

Main highway to city

activities

1 (a) State the meaning of the term 'green belt'. [2]

(b) Which city was the first to adopt a green belt? [1]

2 (a) Copy the following table headings and write in as many possible land uses seen in the rural–urban fringe that you can think of.

Transport	Housing	Industry	Retail	Farming	Urban services	Leisure	Other

(b) Imagine you are a farmer who owns land in the rural–urban fringe. Write an article to your local newspaper outlining the problems you have because of your location and outline your plans for the land. Use your computer to present your work like a newspaper page. [10]

3 State fully two reasons why large retail centres are choosing to locate in the rural–urban fringe. [6]

HIGHER TIER

4 (a) Using OS or aerial maps of Belfast, or another UK city, examine the rural–urban fringe around it. Draw an annotated sketch of the area ensuring you label all the activities you see there that are typical of this zone – use grid references in your answer. [10]

(b) Explain why the use of green belts around cities encourages sustainable urban development. Give at least two separate reasons within your answer. [6]

5 Using the BBC website posted below, write a role-play that covers the issue of green belts. You can then swap them round and read each other's work out loud.

web link & extra resources

http://www.naturenet.net/status/greenbelt.html

http://news.bbc.co.uk/1/hi/england/2698301.stm – this is a good article from 2003 calling for the continuation of green belt areas.

http://website.lineone.net/~greenbelt/index.htm#North – this gives the history of London's green belt and has articles about current issues regarding development in green belts.

CASE STUDY

BELFAST

Functional land-use zones

Belfast displays characteristics of both the Burgess and Hoyt models of urban land-use patterns seen in cities.

The CBD consists mostly of large department stores or chain stores that sell comparison goods, like clothing. There are also shops that sell expensive and specialist goods, e.g. jewellers. Other common land uses are offices of local government departments and private businesses.

Outside the CBD the following characteristics can be seen:

- In the west of the city, i.e. west of the Lagan, we see evidence of sectors (see Figs 12 and 16).

- In the east of the city concentric rings can be seen (see Figs 12 and 16).

- There are some self-sufficient neighbourhoods which were previously villages, e.g. Ballyhackamore and Dundonald, that have become incorporated into Greater Belfast as it has expanded outwards. Other independent neighbourhoods, sometimes called nuclei by geographers, are in the form of out-of-town developments like the Belvoir housing estate.

continues

continues

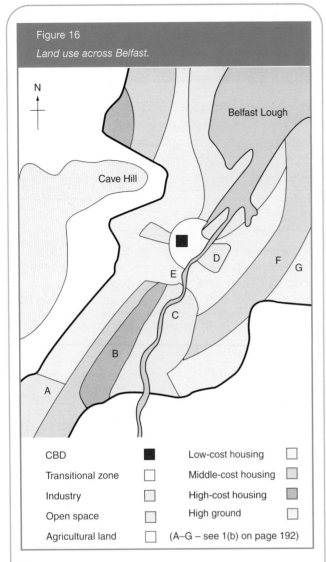

Figure 16
Land use across Belfast.

Belfast Lough

Cave Hill

N

D
E
F
G
C
B
A

CBD	■	Low-cost housing	□
Transitional zone	□	Middle-cost housing	□
Industry	□	High-cost housing	□
Open space	□	High ground	□
Agricultural land	□	(A–G – see 1(b) on page 192)	

Sectors

Evidence of sectors can be seen on Lisburn Road. On one side of the Lisburn Road there is low-density high-quality housing – both detached and semi-detached – similar to that seen on the Malone Road. On the other side there is high-density, lower-quality terraced housing and shops. There is a large sector of similar housing in west Belfast, e.g. the Falls and Shankill Roads. Near to these are sectors of high-quality housing, e.g. Cliftonville and Antrim Roads. A sector of industry also follows the railway and motorway (M2).

Concentric circles

These are more in evidence on the eastern side of the Lagan. The inner part of the city was composed of old established industry, old commercial premises and high-density terraced housing. A lot of this has now been redeveloped, e.g. Lower Newtownards Road and Woodstock

Road. Around this zone land use changes largely in a ring pattern, this is especially clear in the type of housing.

Recent land-use changes in Belfast

Over the last 20–25 years there has been a movement of commercial buildings out of the city centre to the edge of the city. This reflects the movement of people to the suburbs and cheaper rents. Shopping centres like the Abbey Centre have provided a lot of competition for the CBD, especially during the 1970s and 1980s (during the troubles). The result was a decline in the CBD. Revitalisation is now taking place in the city centre, helped by the introduction of pedestrian precincts, late night shopping, better car parking, new shopping centres like Castle Court and all the development in the inner city, e.g. Laganside, which is encouraging people to move back into areas close to the CBD.

Some areas of change include:

- Rural urban fringe – very periphery of the city e.g. Cairnshill, Four Winds, south Belfast. Farmland is being built on.
  New large housing – detached with garages. Wealthier socio-economic groups move in.

- Inner suburbs
  e.g. Sandown Road – off Newtownards Road, east Belfast.
  Some large housing has become run-down. Large houses are redeveloped as apartments or nursing homes.

- Industrial zones of the inner city
  e.g. Laganside.
  Government money helps change land use from industrial to leisure/entertainment and residential, with some service industry.

- Inner City – certain residential areas e.g. Stranmillis/Holylands.
  Undergone gentrification where wealthier people move in and improve the old terraces, pushing original residents out as house prices go up.

Ethnic areas in Belfast

People sometimes move into areas where their neighbourhoods have a similar ethnic background to themselves, creating ethnic zones in the city. A persons ethnicity is made up of their age, nationality, language, race, socio-economic class, gender and culture.

continues

continues

In large urban areas in the USA ethnic zones often include White, Afro-American and Hispanic. In Northern Ireland the zones are divided more often by religion than race.

The two main ethnic cultures have different traditions and culture:

- Catholic;
- Protestant.

Belfast has such strong ethnic zoning that geographers describe it as segregated. Segregation means that members of an ethnic group are not uniformly spread across the residential areas of a city. Some small parts of the city or entire housing estates have a majority of residents that belong to the one tradition or culture.

Two-thirds of people in Belfast live in an area that has over 60% of all residents from either a Protestant or Catholic background.

Is the future segregated?

Some attempts are being made to integrate the two main communities here. There are integrated schools and EMU projects. The police service is trying to recruit religions equally, and employers also have to comply to equal opportunity laws. There are also a few mixed marriages.

We do also have members of ethnic minority groups living in Northern Ireland, with Belfast having the largest concentration.

Figure 18
Percentages of ethnic make-up of Belfast, 2001.

Ethnic group	Percentage
White	98.63
Irish traveller	0.09
Mixed	0.26
Indian	0.16
Pakistani	0.06
Bangladeshi	0.02
Other Asian	0.03
Black Caribbean	0.02
Black African	0.05
Other Black	0.03
Chinese	0.48
Other	0.17

Some neighbourhoods, called wards during a census (a count of the population), also have a concentration of a certain group, e.g. the Botanic ward of Belfast has a large number of young adults.

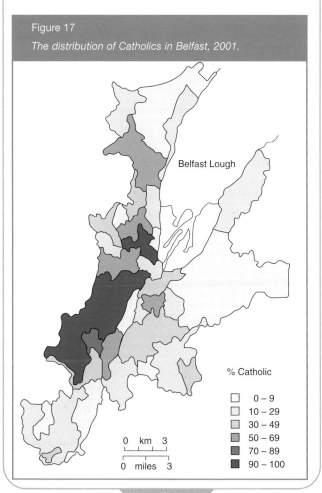

Figure 17
The distribution of Catholics in Belfast, 2001.

Belfast Lough

% Catholic

0 – 9
10 – 29
30 – 49
50 – 69
70 – 89
90 – 100

0 km 3
0 miles 3

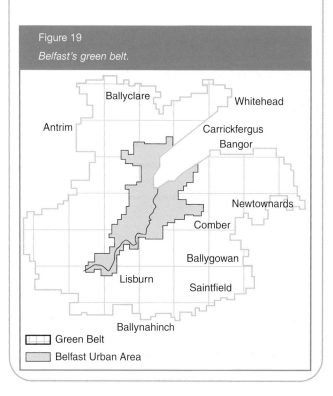

Figure 19
Belfast's green belt.

Ballyclare
Whitehead
Antrim
Carrickfergus
Bangor
Newtownards
Comber
Ballygowan
Lisburn
Saintfield
Ballynahinch

Green Belt
Belfast Urban Area

continues

activities

1 (a) Make a copy of the sketch map showing land use across Belfast (Fig. 16).   [10]

   (b) Using the land-use map of Belfast, locate the areas lettered A to G. Choose your
       answers from the list below.   [7]

   Belmont Ormeau Park Malone Belvoir Park Ballyhackamore
   Dunmurry Industrial Estate Stranmillis

   (c) Name four ethnic groups found in Belfast.   [4]

   (d) Draw a graph to illustrate the information on the ethnic make-up of Belfast.   [5]

2 State fully two ideas you think would help to break down ethnic divisions in Belfast.   [6]

3 Using the property news website and census material online and a street map of
   Belfast, investigate the main residential zones of the city. Create a PowerPoint
   presentation outlining your findings.   [10]

HIGHER TIER

4 (a) "Belfast has definite ethnic areas". Do you agree? Use evidence from the text
       within your answer.

   (b) The Belfast Metropolitan Area Plan (BMAP) recommends that 42,000 new homes
       must be built to accommodate housing demand in Belfast. Describe where you
       would like to see most of these houses being built and justify your answer.   [8]

   (c) Investigate one area of change in Belfast using the internet – for example,
       new city centre shopping complexes, and create a PowerPoint presentation
       on whether you would agree the development should go ahead.   [10]

web link & extra resources

www.belfastcity.gov.uk – this site has lots of
information on Belfast, including the Belfast
Metropolitan Area Plan 2001.

www.bbc.co.uk – search for news stories about the
peace lines and other planning issues about
Belfast.

www.belfasttelegraph.co.uk – use this to look up
reports on developments like Victoria Square in
Belfast.

www.propertynews.com – useful to locate areas of
low-, medium- and high-cost housing in an area.

www.nisra.gov.uk – census information available
about the areas you are investigating in a
settlement.

CITIES IN THE DEVELOPING WORLD

Characteristics of the LEDC city

- Transport is difficult. Some areas like the shanty towns have only dirt tracks. Public transport cannot cope with the demand, and often people are seen almost hanging out of buses.

- There are too few homes available for residents, so many of the new migrants have to build their own accommodation on scrap pieces of land at the edge of the city. Where one shack goes up others follow, and before long a shanty town develops.

- Many have newly established industrial areas, with transnational corporations.

- There are stark contrasts between the rich and poor with low-cost slums springing up next to expensive housing.

Figure 20
A LEDC city.

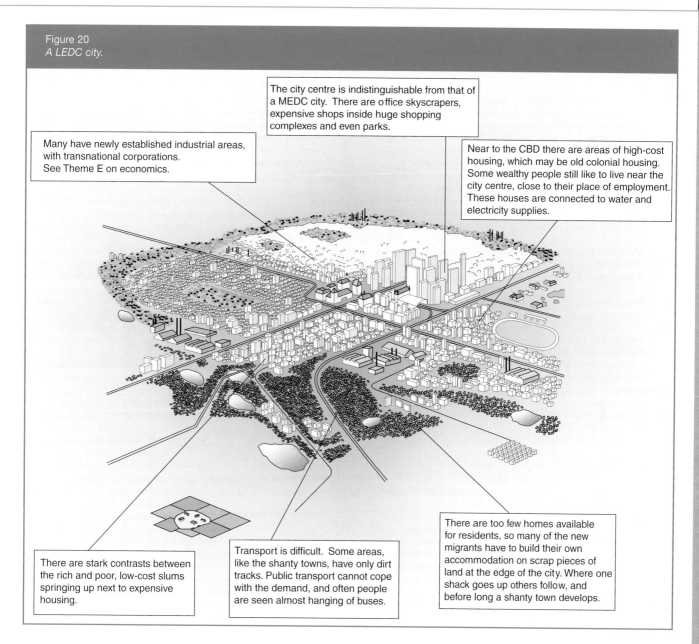

The city centre is indistinguishable from that of a MEDC city. There are office skyscrapers, expensive shops inside huge shopping complexes and even parks.

Many have newly established industrial areas, with transnational corporations. See Theme E on economics.

Near to the CBD there are areas of high-cost housing, which may be old colonial housing. Some wealthy people still like to live near the city centre, close to their place of employment. These houses are connected to water and electricity supplies.

There are stark contrasts between the rich and poor, low-cost slums springing up next to expensive housing.

Transport is difficult. Some areas, like the shanty towns, have only dirt tracks. Public transport cannot cope with the demand, and often people are seen almost hanging of buses.

There are too few homes available for residents, so many of the new migrants have to build their own accommodation on scrap pieces of land at the edge of the city. Where one shack goes up others follow, and before long a shanty town develops.

■ Near to the CBD there are areas of high-cost housing, which may be old colonial housing. Some wealthy people still like to live near the city centre, close to their place of employment. These houses are connected to water and electricity supplies.

■ The city centre is indistinguishable from that of a MEDC city. There are office skyscrapers, expensive shops inside huge shopping complexes, and even parks.

The reality for urban migrants – a shanty town life

The term 'shanty' is an African term for an area of slum housing, often built by the residents from scrap materials, which has few services. They have other names, depending on where you are:

■ Latin America – favela;

■ India – bustee;

■ French-speaking Africa – bidonvilles.

What are the characteristics of shanties?

■ They contain self-built housing, in other words, the people who live there build their own shack.

■ The land is illegally occupied, so the government views them as squatters.

■ Housing is poorly constructed, without proper foundations and from a variety of materials.

■ There are few services, like clean water supplies, sewers for waste and electricity, and on a high level, with few schools, medical centres and public transport connections.

Figure 21
A shanty town.

■ Most of the residents are poor, and have large families – the average family size in the shanties of Kenya is five.

■ As there is no planned street pattern, the housing is haphazard, although established shanties, like Kibera in Nairobi have quarters where certain jobs are concentrated.

Where do we find shanty towns in the LEDC city?

■ Areas of unwanted land – these may be steep slopes prone to landslides like in Rio de Janeiro, or on land next to rivers that flood, like some bustees in Mumbai.

■ On the edge of the city, where larger tracts of land can are occupied by new arrivals from the countryside. They often follow main roads or railway lines.

In some LEDC cities, there are also pavement dwellers, who have built shacks along the side of the road on the pavement. These shelters are illegal, but are often tolerated as people need to live near their work. In Mumbai, some of these shelters have been lived in for almost a decade, and they have even persuaded the local postman to deliver letters to them!

activities

1 (a) Complete the following sentences about the LEDC city. Choose your answers from the word list below.

Crowded   TNCs   close   difficult
poor   many

(a) Transport is _____.

(b) Streets are _____.

(c) Large numbers of people live in _____ housing in shanty towns.

(d) Formal industrial areas have many _____.

(e) High-class housing is often _____ to the CBD.

(f) The CBD has _____ skyscrapers, just like a MEDC city. [6]

2 Describe two possible locations shanty towns occupy in a LEDC city. [4]

3 (a) Write a paragraph describing conditions in a shanty town. [6]

(b) State fully one way you think shanty towns could be changed for the better. [3]

HIGHER TIER

4 Describe and explain two differences between land use in a LEDC city and that predicted in the concentric ring model based on MEDC cities. [8]

CASE STUDY

SÃO PAULO, BRAZIL

Brazil is very industrialised for a LEDC, and its population distribution reflects this. A total of 80% live in urban areas – and São Paulo is the biggest of all the Brazilian cities. São Paulo is located on Brazil's south-east coast. It is has population of approximately 25 million, this is larger than Rio (10 million) and Recife (3 million). It is one of Brazil's 23 millionaire cities.

São Paulo is on a plateau, 70 km inland and at 730 m above sea level. This gives it a pleasant cool climate. The rich terra rossa soil, from the weathered lava underneath it, is very good for farming and is mostly used to grow coffee. This attracted people, and goods could easily be imported and exported using the Parana River system.

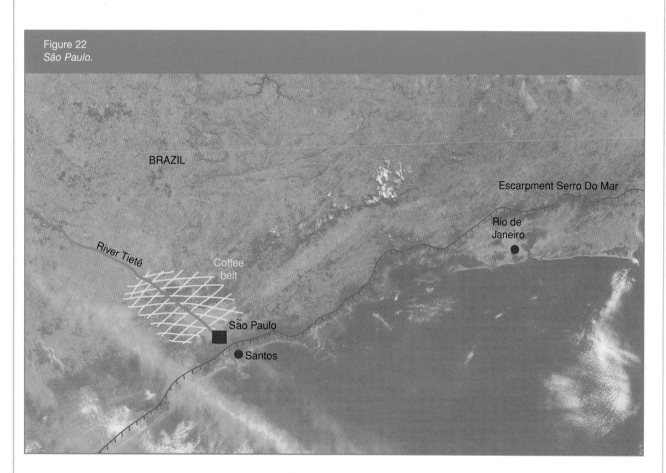

Figure 22
São Paulo.

The population of São Paulo has grown for a number of reasons. Natural increase is one reason for its growth (this is when the birth rate is higher than the death rate). The population has also grown as the result of urbanisation. The has been caused by rural to urban migration. Millions of people have migrated from Brazil's rural areas to São Paulo. In São Paulo 65% of urban growth is a result of migration. This is caused by a variety of push and pull factors.

The growth of São Paulo began with industrialisation. From the 1950s onwards people were being attracted to the city by the prospect of finding a job in one of São Paulo's factories – at this time manufacturing, especially car manufacture, was overtaking coffee as the main income earner for the city. In the 1970s there was massive out-migration from the countryside as the development of HEP schemes and mechanisation of farming meant that over 4 million Brazilians could no longer earn a living there.

continues

São Paulo is the third largest city in the world.

São Paulo has more tower blocks and a denser population than New York!

Land-use zones

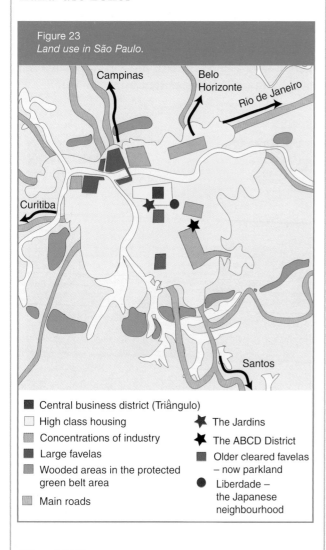

Figure 23
Land use in São Paulo.

Campinas

Belo Horizonte

Rio de Janeiro

Curitiba

Santos

■ Central business district (Triângulo)
□ High class housing
■ Concentrations of industry
■ Large favelas
■ Wooded areas in the protected green belt area
■ Main roads

★ The Jardins
★ The ABCD District
■ Older cleared favelas – now parkland
● Liberdade – the Japanese neighbourhood

The CBD

The central business district of São Paulo, as in Belfast, is right in the middle of the city. It is locally known as the Triângulo. It is the most accessible part of São Paulo, being the meeting point of the main roads and rail links. It has the greatest concentration of retail and commerce, and resembles the CBD of any MEDC city, with wide avenues and skyscrapers.

Older parts in the centre of São Paulo are also being regenerated. The Anhangabaú Valley, the old heart of the city, was becoming swamped by traffic in the 1980s. Redevelopment began in 1991

Figure 24
The Banespa and Martinell. Buildings in São Paulo's CBD. The Martinell Building (the reddish Victorian building on the left) is Latin America's oldest skyscraper.

involving the creation of an area of open parkland, now used by 1.5 million people each day who cross the centre of the city – that's the equivalent to the whole of the population of Northern Ireland each day! Traffic problems were tackled by the construction of 570 metres of underground tunnels to bypass this section of the city. Many of the old historical buildings were also renovated.

Industrial area

There is informal industry throughout the city, as people make and sell things from the favelas and come into the CBD to provide services. However, the main concentration of formal industry is in the part of São Paulo known as the ABCD complex. This name comes from the districts it is made up from – Santo André, São Bernardo, São Caetano and Diadema. Here we find many of the main car factories. Companies that chose São Paulo as their base in South America include Ford, General Motors and Volkswagen. It is located to the south-east of the CBD on the main infrastructure links to the port of Santos. It is also well connected to the main city airport in the south of the city.

Residential zones: high-cost housing

Because São Paulo has grown so quickly, the evidence of circles of similar land use around the CBD is minimal. Directly next to the CBD is the only obvious concentric circle which consists of high-class housing. Many of the houses here are old colonial stately houses, making them large and so expensive. The middle classes favour this area

continues

continues

as they are close to the CBD with the main concentration of shops and amenities. There are newer blocks of flats in this area and walled Californian-style detached housing with their own swimming pool and garden. The residents will have at least one maid to help with the household tasks and their children get educated in Private schools, meaning they have good career prospects.

Figure 25
The Jardins district.

One such area is the Jardins. This is an affluent planned zone that was created in 1915. As the name suggests (Jardin being French for garden) there is a lot of greenery here – 50 m² of green area per capita, which is equivalent to most MEDC cities. Most of the housing here is in the form of exclusive apartment buildings, which have designer clothes shops and expensive restaurants on the ground floor.

Residential zones: low-cost housing

The rapid growth of São Paulo's population has led to a severe shortage of housing. Millions of people have been forced to construct their own

homes from scrap materials such as wood, corrugated iron and metals. These areas of temporary accommodation are known as favelas in Brazil. Many of these are on the outer edge of the city, and now people can spend up to 4 hours a day commuting to and from work in the CBD. This pattern of residential land use is like the reverse of Burgess model – in São Paulo, much of the high-class housing is near the CBD and the lowest cost housing is on the periphery of the city. The housing in favelas is a reflection of socio-economic status – the better off they are, the better the housing they can afford.

Figure 26
Location of favelas in the municipality, 1987.

	Number	Percentage
On the banks of streams and rivers	783	49.2
Areas subject to flooding	512	32.2
On steep slopes	466	29.3
In areas of accentuated erosion	385	24.2
On rubbish tips	30	1.9
Beside major roads	40	2.5
Beside railways	25	1.6

The conditions associated with favelas are very poor. Often families have to share one tap, there is no sewerage provision, disease is common and many people are not employed in the formal sector, i.e. there is a high rate of underemployment.

Reasons why people move into the favelas:
- Low price of accommodation;
- Often close to factories on the edge of the city;
- No rates to pay to the government.

The location of some of São Paulo's favelas, together with the poor construction of many of the buildings, makes them vulnerable to damage during natural disasters.

One recent tragic example of the uncertain nature of life in the shanty towns of São Paulo happened on Tuesday 29 February 2000. Sadly 13 people died after a mudslide collapsed and buried their

continues

continues

Figure 27
The Paraisopolis favela.

flimsy wooden shacks built at the side of a hill in São Paulo's southern suburbs.

According to a census carried out in 1999, 30,000 people live in the Paraisopolis favela – the largest favela in the city of São Paulo. However, the locals put the number nearer to 60,000.

Situated within the upper-class neighbourhood of Morumbi, the Paraisopolis favela is like a small town inside the mega-city of São Paulo. This is typical of land use in many LEDC cities, the juxtaposition of rich and poor, the two faces of the city. Many of the houses in Paraisopolis are made from bricks, which at least gives them a stable frame. In many of the other facelas in São Paulo, the houses are made from wood. However, there is no proper sewage system.

Houses are bought and sold in this long established favela. The average cost of a small two-bedroom house here is just over £10,000. This compares favourably with areas of the city that have basic services like sewage and street lights. In those areas the same size house would cost about £25,000.

Figure 28
Liberadade, the Japanese area of São Paulo.

Are there ethnic areas in São Paulo?

São Paulo is much more culturally diverse than Belfast, and people often classify their ethnic origin by their race.

Over 1 million Japanese live in São Paulo City alone, making it one of the largest Japanese communities outside of Japan. The Japanese are not the only immigrant group that makes of São Paulo one of the world's most cosmopolitan cities. São Paulo has inhabitants from over 70 different nationalities. Over 3 million Italians, 1 million Arabics, 1 million Portuguese and 1 million eastern Europeans also live there. There are also indigenous peoples (original inhabitants) of Brazil, many of whom are represented in São Paulo's population.

Green areas and green belts

In a city so full of people and so pressurised for new housing and development, you might think that São Paulo would have almost no green areas, and that no protection exists for the countryside surrounding it. That is not the case. The wooded areas around the city are in protected green belt land and close to the CBD there is a large park area called Ibirapuera Park.

continues

activities

1 (a) List two advantages of the site of São Paulo. [2]

(b) São Paulo is growing rapidly. State two pull factors attracting people to the city. [2]

(c) What is the local name for the CBD of São Paulo? [1]

(d) Describe urban change made in two different zones of São Paulo. [6]

2 (a) What kind of housing is found in the Jardin district of São Paulo? [2]

(b) Outline the lifestyle of the residents of this district.

3 (a) State three reasons why people are moving into the shanty town areas of São Paulo. [6]

(b) With reference to Fig. 26, describe the location of shanty towns in São Paulo.

(c) Name three ethnic groups found in São Paulo.

HIGHER TIER

4 Compile a short PowerPoint presentation on São Paulo, ensuring it includes the following information: map of location of the city, characteristics of the CBD, high-cost housing zone, low-cost housing zone, green belt and one area where change is occurring. [15]

web link & extra resources

http://landsat7.usgs.gov/index.php – image of São Paulo.

www.bbc.co.uk – search under the world news section for São Paulo to keep up to date with the city.

www.zonalatina.com/Zldata124.htm – the city's history can be found here.

hometown.aol.com/pochetti5/sampa-brazil.html – this site has amazing photographs of São Paulo.

www.bennett.karoo.net/topics/urbanproblsledcs.html#saopaulo.

CHANGE IN URBAN AREAS

Change in MEDC cities

Housing

Most inner-city areas developed along with industry in the 19th century. As industry grew so did demand for workers and rural–urban migration took place. At that time people wanted to live as close as possible to their place of work as possible due to the lack of public or private transport.

Nineteenth-century housing was high density, often back-to-back terraces. By the 1960s, when many houses lacked basic amenities, large-scale slum clearances led to the building of high-rise flats, but these flats created as many problems as they solved.

Today many older terraced housing and high-rise blocks have been abandoned, boarded up or vandalised. Some have been improved (urban renewal) and others have been demolished or replaced with new houses of lower density (urban redevelopment).

Industry

Inner cities were characterised by large factories built on land adjacent to canals and railways. Many of these have been forced to close due to lack of space for expansion, the narrow congested roads and movement of the main business abroad.

Some factories have been left empty and others have been pulled down, leaving large areas of derelict land.

In some cities there have been major schemes to regenerate these areas, e.g. Laganside in Belfast and the Docklands of London.

Retail

Many of the traditional corner shops have been forced to close due to competition from city centre shops and shopping centres. Major food retailers are choosing locations away from the CBD to allow for large superstores and free car parking for their customers.

Consequences of change in urban areas

On people:

- Traffic congestion means longer journey times;
- Increased volume of traffic means greater risk of accidents;
- Noise and air pollution affects health;
- High land values mean high house prices;
- Ethnic tension in inner cities;
- Lack of jobs in inner cities as industries close down.

On environment:

- More traffic means more noise and air pollution;
- Land around the city edge is used for building houses, increasing the distance between the CBD and the countryside;
- Areas of derelict land in the inner cities where industry has closed down.

Change in LEDC cities

As many of these cities are in a wealth-accumulating era, change is very rapid at present. One area of these cities that is changing particularly fast is the lowest cost housing in the shanty towns.

The favelas are sometimes seen as a problem in the way of other developments. The management response here from the government is usually the clearance of the shanty area, which they can do as the residents often have neither bought nor rent the land they occupy, in other words they are there illegally. Read the following report taken from a São Paulo newspaper in 1997.

---

250 military police accompanied by court officials initiated the expulsion of 1800 families (approximately 7200 people) from a shanty town (favela) in Santo Andre in the Greater São Paulo area on May 7. The families being dislodged from the favela are homeless and had occupied a 209,000 m^2 area which had been purchased by the previous city council.

Many of the shanty town dwellings were destroyed by tractors; others were burned by their owners. The clearance of the area is expected to take five days.

Source: www.oneworld.org.sejup/226.htm

This area was cleared as part of the Cingupura Project in São Paulo. The problem is that even though the new flats are cheap, they are still too expensive for many of the favela dwellers, who then have to start the search for a new scrap of land to build on.

Another response to the need for change in the favelas is by providing self-help schemes to create basic, but affordable, housing for the poor. These initiatives are funded by the World Bank as well as voluntary aid organisations.

Such schemes in the poorest peripheria of São Paulo have improved services such as water supply, sewage and electricity.

These schemes are cheap and cultivate a feeling of community spirit as the basic housing is often constructed by local residents. This enables costs to be kept low. Also, it does not involve the forced removal of large numbers of residents like the larger scale improvements seen in São Paulo, which involve slum clearances.

There are some problems, however. People must be willing to work together and organise themselves into co-operatives to get the work done. Single-storey constructions, as shown in Fig. 29, or even two-storey ones are not creating a high enough population density to be sustainable, despite being close together. They make cities like São Paulo low rise and spreading, which makes it expensive to provide comprehensive public transport connections and other needed services.

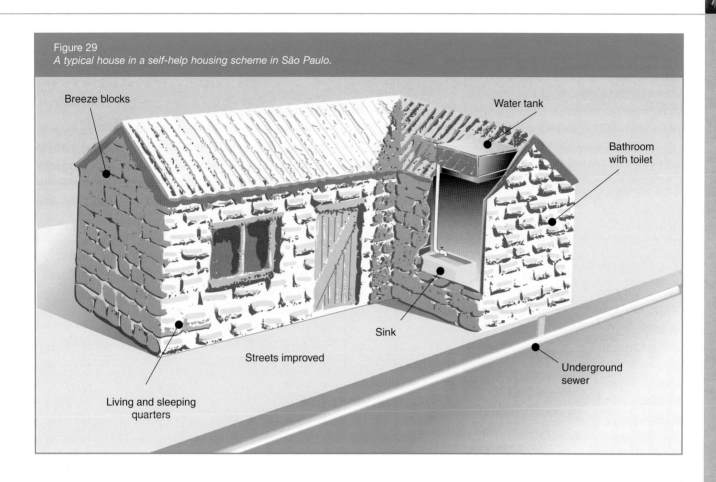

Figure 29
A typical house in a self-help housing scheme in São Paulo.

Breeze blocks

Water tank

Bathroom with toilet

Sink

Streets improved

Underground sewer

Living and sleeping quarters

activities

1 (a) State fully three reasons why cities are always changing. [9]

   (b) Choose two of the following changing areas of a city as discussed in the text. Describe and explain the changes for that land use.

     Housing   Retail   Industry [6]

2 Research urban change in your local town or city. Compile a report for your teacher that outlines the change, and the groups of people it might affect. Make up a questionnaire to ask other pupils in your class to assess their opinions on the proposed change. [10]

HIGHER TIER

3 (a) Describe two methods used in São Paulo to change favelas (shanty towns). [6]

   (b) Which method do you think is appropriate? Justify your answer. [5]

4 (a) Draw a simple labelled sketch map to show the zones of a city you have studied in a LEDC. [5]

   (b) Select one zone in this city and explain how and why it has changed over time. [6]

UNIT THREE
Planning Sustainability for Urban Environments

Planning is the carrying out of a programme of work, such as building a new estate, or protecting a building, using an agreed design or set of guidelines.

Sustainable development of cities means that they should grow in a way that meets our needs today, without making it difficult for future generations to meet their own needs.

The characteristics of a sustainable settlement:

- Provides enough secure jobs.
- Makes available a mix of land uses throughout the city.
- Provides housing, of different types and costs.
- Minimises the use of non-renewable resources – reduce, reuse, recycle.
- Provision of medical support for all who need it.
- Local community is involved in decision making about the city.
- There is access to education to all equally.
- Historical buildings are conserved.
- Traffic flows well, with most people using public transport.

In Northern Ireland we have been specifically concerned with sustainable development since 1992, when the UK signed up to an action plan, called Agenda 21, during a UN summit on Environment and Development. Agenda 21 directs us towards sustainable development in the 21st century, identifying the role and actions for all sectors of society – government, businesses, schools, women, ethnic groups and local authorities are all given roles to play.

Today, most of the urban development in Belfast is happening in two of the land-use zones identified by Burgess in his ring model:

- On the edge of the high-cost residential zone;
- In the transitional zone.

Change in the edge of the high-cost residential zone

This area is under pressure, because for developers it is the cheapest place to build new housing as greenfield sites do not need expensive preparation for building and the tax cost is less. However, the government sees the need to contain the spread of the city (see the section on green belts on page 187), so more and more of the current building in cities is happening on brown field sites in inner-city areas, also known as the transitional zone.

Primrose Hill, on the Saintfield Road, is on the edge of the high-cost residential zone. It has a mix of housing types to encourage a socio-economic mix in the estate. The developers left land for the building of a community facility – a large church. There are also trees planted with the development.

Change in the transitional zone

In the transitional zone the inner city is being regenerated – it is changing from the image of a run-down, dirty area into an area of exciting modern places to live and work. Just like an ageing pop star, the inner city is reinventing itself.

Regeneration

Individuals buying and/or improving an old house can regenerate an inner-city area.

They might add central heating, new kitchens and bathrooms, or even extend the building. This increases its value, and encourages slightly wealthier buyers when the house is sold on. When enough people do this in an area of old terraced housing, such as Stranmillis, instead of having the traditional lower paid residents, the area attracts well-paid professionals. We call this **gentrification**. Look at the photographs below. Which area looks gentrified and why?

Regeneration may involve large organisations, such as the Laganside Corporation. Set up in 1989, this organisation put forward a plan to change large areas of inner city Belfast next to the River Lagan. The once vital industrial area of docks there, which made Belfast one of the most important cities in the UK during Victorian times, was run down and polluted in the 1980s. This happened because:

- Modern industry preferred more attractive sites to locate on the **urban fringe**.

- The docks moved down-river in the 1960s to deep-water locations that the new large freight ships needed.

- As the river is tidal, mudflats were exposed at low tide. These were ugly, often covered in rubbish and smelt awful in warm weather.

The aim of the Laganside Corporation was to improve the quality of the River Lagan and redevelop land along the banks of the river to encourage a rebirth of this long neglected area. To start with they planned the redevelopment of 140 hectares (ha) that runs along a 4.8 km stretch of the River Lagan. Sustainability was very important to the planners involved, and in the next few pages we will see how they completed this transformation.

web link & extra resources

www.belfastcity.gov.uk – provides free booklets and leaflets.

www.gdrc.org/uem/ – a virtual library on the urban environment.

www.bbc.co.uk – search under news for urban planning.

activities

1 State the meaning of the term 'sustainable development'. [2]

2 Using the information in Fig. 30 state fully two ways these characteristics of a sustainable settlement mean that people can continue to have a good quality of life in a city. [6]

3 Describe two ways the inner city may be regenerated. [4]

HIGHER TIER

4 Suggest two advantages and two disadvantages of gentrification in inner city areas of Belfast. [8]

LAGANSIDE – A LOCAL PLANNING INITIATIVE

Since the 1990s, the Laganside Corporation has successfully regenerated just over 200 ha of Belfast's inner city along the banks of the Lagan. Go to the website www.laganside.com to see an interactive map of the key Laganside sites as mentioned below.

Laganside has become a focus for regeneration by:

- Providing much needed investment in the roads, railways and waterways (infrastructure).
- Creating an improved environment.
- Serving new investment in housing, jobs, leisure and recreation.

The River Lagan

In 1994 the Lagan Weir was completed. The weir looks most impressive at night when it is lit up (see Fig. 31 below). It cost £14 million and was one of the most complex engineering projects seen in Belfast's history. It now controls the tidal reach of the Lagan, ensuring water levels in the river remain high, even when the tide is out, so covering the unsightly and smelly mud banks(see Fig. 32 below). The weir also gives Belfast protection against flooding.

The roads and railway

Opened in 1995, the M3 Lagan Road Bridge links the M1 and M2 with the Sydenham by-pass and Newtownards Road. This road, together with the cross-harbour rail link between Larne and Belfast, has helped reduce the volume of traffic in the city centre.

The built environment and green spaces

A key feature to all the Laganside sites is the mix of building types and land use – in other words they are designed to be areas where people can live, work and play. Looking at Lanyon Place, covering 6 ha, and once an important market place in the city, this concept of combining building types is clear.

The Waterfront Hall

A major new public building for recreation, combining modern architectural design with a traditional slant – the roof will eventually 'weather' to be a bright green like many of Belfast's public buildings.

The Hilton Hotel

A major investment from a well-known luxury hotel chain. It is Belfast's first five-star hotel, built at a cost of £21 million.

Figure 31
The Lagan Weir at night.

Figure 32
The site before the Weir was built.

Multi-storey office buildings

The most recognisable being the British Telecom Tower. Many of these buildings have been commissioned by international architects, such as Neils Torp (Norwegian), who designed the predominantly glass office building opposite the Waterfront Hall.

Gregg's Quay

Directly across from Lanyon Place are 32 apartments that were completed in 1999. The award-winning design of Gregg's Quay uses brick, stone and glass with spectacular effect.

However, not all the building work in Laganside is new. Some of the old buildings have been conserved. Careful restoration work on the Gasworks has allowed it to be reopened to the public after almost 200 years. St George's Market has also been conserved, and is once again a successful market area.

The improvement of the non-built environment, by the creation of green areas and public spaces, is one of the most notable changes the regeneration of Laganside has brought with it. The Gasworks site was once so contaminated that even weeds found it hard to grow. Today, after reclamation, the buildings almost take a backseat to the public parkland and pathways, full of art and sculpture that reflect the history of the location.

Amenities

In the late 1980s Laganside had very few amenities. Now, through regeneration, there are many pleasant and useful features (amenities). Buildings such as St George's Market, the Odyssey and the Waterfront Hall draw visitors from all over Northern Ireland to attend the public events they hold. They also provide local people with leisure facilities and job opportunities – over 600 jobs have so far been created at the Odyssey alone. The Cathedral Quarter is receiving new investment to bring a cosmopolitan feel to Laganside. There is a developing café society, and the Cathedral Quarter Arts Festival attracts thousands of visitors. Belfast's only circus school has taken up residence in one of the buildings here. A new network of public walkways and cycle paths has been set up. These stretch outwards from Belfast's city centre through the Lagan Valley Regional Park towards Lisburn. The Millennium National Cycle Route allows cyclists to go along the Lagan and on to Lough Neagh. These pathways, together with pontoons and fishing platforms, have greatly improved public access to the River Lagan.

Figure 33
The Gasworks, before and after renovation.

Housing

A large proportion of the housing built at Laganside is composed of apartments. Being so close to the CBD, bid-rents are high. By stacking buildings upwards, in apartment blocks, developers can make enough money from the sales to cover the cost of the land. Almost 400 apartments have been constructed across the whole of Laganside. Most of these are in the Mays Meadow site. Many of these new apartments are much more expensive than the traditional terraced housing found next to Laganside, in areas like the Short Strand. To encourage a mix of social groups, there have been attempts to mix housing types and prices. For example, 48 of the apartments in Mays Meadow were offered by Belfast Improved Housing to people of modest means, and at Ravenhill Reach there are 36 housing units, controlled by a housing association, which provide housing for mature singles. Public housing has also been included to reduce social exclusion.

Employment

The Laganside Corporation has been very successful at attracting investment to its sites, with almost £700 million being invested since the early 1990s. Altogether over 8000 jobs have been created. These jobs are very different from the types of jobs that the area offered in 1900.

The heavy industry that once dominated Laganside has given way to modern hi-tech industries. Take Clarendon Dock as an example. This 12 ha site, once known as Ritchies Dry Dock, was where boats were brought for repairs. Today it is a successful business park with many international companies, including Prudential PLC (whose call centre employs over 500 staff), Tesco and Allied Dunbar. It is also home to the Northern Ireland Council for the Curriculum, Examinations and Assessment (CCEA). Clarendon Dock has been attractively landscaped with natural stone and tree-lined boulevards. Its history has been preserved by the restoration of the dry docks and its public spaces have been opened up around Barrow Square, where music festivals are held near the Rotterdam Bar and Pat's Bar.

Recycling and reusing

Some materials from the original area were reused as keystones in the new buildings.

Benefits and problems to local communities

In 1997 Laganside adopted a new 'Community Strategy' to ensure that local communities would also benefit from the changes brought about by the regeneration of Laganside. There are three strands to this policy:

- Keeping local people informed of development plans and asking for their opinions on the proposed changes.
- Working with individuals and the community to help them retrain, so they can apply for jobs being offered by the hi-tech firms.
- Creating and/or funding social activities for all to enjoy.

The quarterly news-sheet produced by the Laganside Corporation, called Laganlines is distributed to 180,000 homes in Belfast, and keeps people informed about events and new developments. Links have been established with community groups closer to the Lagan, and they can have an input into many issues, from providing secure fencing along the river, to making a playground or public piece of art.

Training schemes have been set up through the Training and Employment Agency in conjunction with some of the new firms, like the Hard Rock Café and Halifax, to train local people for the jobs that are now available. There has also been a few special programmes, such as Millennium Leap, set up in the Cathedral Quarter, which trains single parents to work within the hi-tech IT sector.

The Corporation also provides free water safety classes for schools and community groups. In addition, it also gives grants to allow themed parties and events to be held.

All this effort shows that Laganside Corporation recognises that the support and inclusion of local communities is vital to the holistic regeneration of the area.

Is all this change good?

Sean Flynn – Community Development Officer

The community of the Markets Area, who I represent, are angered by the urban regeneration. The sense of community that was once so obvious here, is disappearing. The history of this area, as the 'bread basket of Belfast', full of bakeries and vibrant markets, is being ignored. This means that we are facing an inner city community with no attachments to the history and traditions, a quasi-community with no soul. We are kept at arm's length from the new business, who employ people from the suburbs, rather than locally. We can't sustain anything here if we do so in ignorance of the people and their past.

Mrs Strong – resident of Gregg's Quay

It is a pleasure to look out at the river Lagan every day. There are lovely tree-lined walks and at night the lights under the bridges make the river look beautiful. Being retired, it's great to be able to walk to the Waterfront Hall, the Odyssey and the city centre. We are delighted to be Laganside residents.

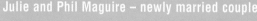

Julie and Phil Maguire – newly married couple

We both lived near Short Strand and have been together since High School. We got married last year, and wanted to live near our parents, but not in a terraced house, since Julie was injured in a car accident and can't manage stairs well. One of the new apartments would have been perfect, but we couldn't afford to buy one. Julie stays at home and I'm a security guard, so we don't have a large family income. The apartments cost about £30,000 more than we could ever afford. We ended up moving out of Belfast altogether and really miss the 'craic' we had with our neighbours and mates.

Jon Norris – manager of Hard Rock Café

The excitement generated by the opening of Odyssey confirmed that this was the right place for Ireland's first Hard Rock Café. We pride ourselves in the staff training programme made possible by Laganside,.it allows us to bring in staff that can continue to offer our customers a quality service and experience.

web link & extra resources

www.laganside.com

www.bbc.co.uk/northernireland/education/ks3geog raphy – there is a good interactive exercise on Laganside available.

activities

1 (a) State the meaning of the term 'urban regeneration'. [2]

(b) Create a table of benefits and problems the Laganside regeneration scheme has brought to local people. Include at least three of each. [12]

2 (a) Name one urban planning initiative you have studied. [1]

(b) Look at the list below, showing aspects of the city that planning often changes.

Green spaces   Services
Housing   Employment

Chose any two aspects and state fully how your named planning initiative has changed them. [6]

(c) Imagine you are a local resident in the area that has been changed. Explain what the planning initiative has meant for you. Include one benefit and one problem in your answer. [6]

HIGHER TIER

3 "The Laganside Corporation has enriched Belfast's inner city, creating sustainable urban regeneration." Using evidence from the previous pages, write a paragraph either agreeing or disagreeing with this statement. Include at least three arguments within your answer. [9]

WASTE MANAGEMENT – RECYCLING AND REUSE

A short walk or drive around your local area in Northern Ireland will demonstrate the problem we have with waste disposal. Inner-city areas in particular have higher levels of litter.

The following results show a comparison between one street in an inner-city area and one in an outer suburb of Belfast. These streets were examined on the same day, a Friday and in the same month, June, and cover the same distance, 500 m.

Type of litter	Inner city	Outer suburb
Crisp packets	15	1
Bottles	7	0
Plastic carrier bags	13	1
Cigarette butts	56	3
Sweet wrappers	24	1
Fast-food boxes and wrappings	5	0
Papers, magazines and leaflets	2	0
Other litter	35	1

As a whole, Northern Ireland has a **waste management** problem. In 1999 Northern Ireland produced 876,500 tonnes of household waste, and the amount is rising at 2% per annum. As a society with a high level of consumption and high GNP, we are wealthy enough to see waste as something that must be disposed of. In many LEDCs it is viewed more as a resource, for example street kids in São Paulo sort through rubbish tips and salvage all the recyclable materials to sell on, and in India they do not throw out edible garbage, instead it is left outside the house for wandering cows to eat – this is part of their religious duties.

What do we throw out?

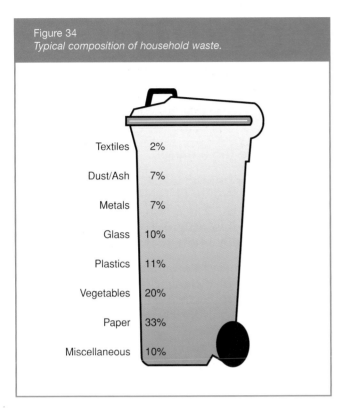

Figure 34
Typical composition of household waste.

Textiles	2%
Dust/Ash	7%
Metals	7%
Glass	10%
Plastics	11%
Vegetables	20%
Paper	33%
Miscellaneous	10%

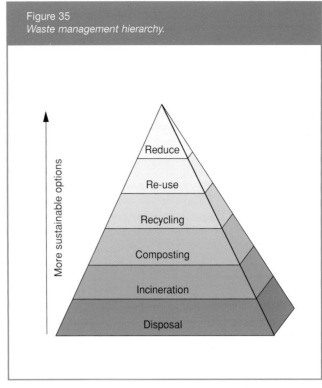

Figure 35
Waste management hierarchy.

More sustainable options

Reduce
Re-use
Recycling
Composting
Incineration
Disposal

In Northern Ireland, we have a poor track record of dealing sustainably with our waste. In 2000 95% of waste was disposed in landfill sites, and only 5% was recycled. Under new EU legislation the UK, including Northern Ireland, will have to ensure that less than a third of its waste is sent for burial in landfill sites by 2020.

Disposing waste by landfill involves burying the waste in a large hole in the ground. Although cheap, there are problems of finding suitable sites, transportation of waste, and air and water pollution caused as the rubbish rots.

To try and make our waste management more sustainable the government has been running a campaign here called Wake up to Waste. It encourages us to implement the 3Rs of waste disposal – Reduce, Reuse and Recycle.

Strategies adopted by central government in 2002 to encouraging people to change current waste management are:

- Landfill tax will increase by £3 a metric ton from 2005–6.

- More local councils will try out rewards for households recycling waste and penalties for those who do nothing.

- Ireland has introduced a tax on plastic bags, charging about 10p (15 cents) for each bag.

- Introduce a doorstep recycling collection service for every household in Britain.

An example of a recent campaign in Northern Ireland against the rash of nappies in rubbish bins.

Real nappies don't cost the Earth

It's Real Nappy Week (7–13 April 2003) and 'Wake up to Waste' is highlighting the environmental impact of 'disposable' or single use nappies and raising public awareness of the availability of real and convenient nappies which 'Don't Cost The Earth'.

Approximately 10 million single-use nappies are dumped in Northern Ireland every month and each one takes on average 500 years to decompose after they have been buried in a landfill site. For every household that has a child wearing nappies, 50% of their waste is made up of single-use nappies.

Councils throughout the province spend a further £1.4 million each year to dispose of these single-use nappies. This then has an adverse affect on the ratepayer who has to pick up the tab for this convenience.

Switching to reusable nappies is clearly desirable. However, in reality we realise that the Northern Ireland consumer should have a choice and what we would encourage is for parents to initially try using both single-use and reusable nappies.

activities

1   (a) Using a spreadsheet package, create a graph to illustrate the differences in litter between streets in an inner-city area and those in the outer suburbs. Use your own figures or those given in the table earlier.   [10]

    (b) Describe and explain the trend shown by your graph.   [8]

2   (a) State fully why our waste disposal is currently unsustainable.   [3]

    (b) Name and describe one sustainable method of dealing with waste and give two examples of how you could carry your chosen method out.   [5]

3   How is the government trying to tackle the problem of waste disposal? Give at least two strategies within your answer.   [6]

HIGHER TIER

4   Which strategy of waste management do you think is most likely to make a difference. Explain your answer fully.   [7]

web link & extra resources

www.sniponline.org/ – a great websites on all things sustainable, including waste disposal.

www3.iclei.org/egpis/#keyinfo – this European site examining sustainability projects all over the continent contains examples of good practice. It also comes in German and French, so test your language skills.

www.wakeuptowaste.org/news/ – one of the best local websites on waste disposal in Northern Ireland.

www.sustainabledc.org/su00014.htm – recommended for Higher Level only.

MEASURES TO CONTROL TRAFFIC

Traffic in general, and cars in particular, is the cause of many problems in urban areas. Traffic levels in urban areas has dramatically increased in recent years. For example, in 1996 there were 23 million cars on Britain's roads, compared to only 15 million in 1980.

Within a day there are patterns in the traffic flow. There are peaks and troughs of flow on certain roads at particular times.

When the road network cannot cope with the demands of traffic congestion occurs. Serious congestion, where traffic is at a standstill for prolonged periods of time, is known as gridlock

Why has traffic in urban areas increased?

- Greater affluence leading to increased car ownership.
- More people commute to work or travel to city centres for shopping/entertainment.
- Reduction in public transport, meaning more private cars.
- Increased road freight, e.g. delivery lorries.

What are the damaging effects of increased traffic in urban areas?

1 Environment
  - Air pollution from exhaust emissions – CO, NO_2, lead, low level ozone and particulates.
  - Noise pollution from cars, lorries and buses.
  - Visual pollution of motorways and car parks.

2 Economy
  - Congestion, especially at peak times, e.g. in London in 1996 the average speed of traffic was 20 km/hr (12 mph) – the same as it was in 1900!
  - Time and money wasted as employees sit in gridlock.
  - Cost of building and maintaining roads.
  - Cost of petrol/diesel.

3 People
  - Danger of accidents and increase in stress for drivers and pedestrians.

4 Buildings
  - Destruction of property to make way for new roads and car parks.
  - Damage to buildings caused by traffic vibrations.

How can transport systems be managed to reduce the damage?

- Exclude, reduce or accommodate traffic by schemes such as park-and-ride and urban motorways.
- Try to reduce pollution, especially from vehicle exhausts.
- Improve public transport, e.g. supertrams – rapid transport systems.

Investigate what traffic calming schemes are in operation in your local area.

CASE STUDY

FREIBURG, GERMANY

The city of Freiburg has a population of 200,000. It is located in the south-west of Germany, next to the Black Forest. It was originally a medieval city, so is encircled by a wall.

In the 1950s the decision to preserve the historical city centre prevailed against proposals to establish an car-centred urban core.

Today, after three decades of 'green' transport policies, it is an outstanding example of best practice in sustainable development, integrating land use and transport planning.

Planning measures implemented

1 Traffic calming

  The city centre was made a pedestrian zone in 1973.

  Natural stone surfaces were added to roads.

  In 1990, a 30 km/hr speed limit was introduced to residential areas.

  Between 1976 and 1992 the percentage of car use fell from 60% to 46%.

continues

2 Continuous upgrading of public transport

Rather than buses, in 1969 the city council opted to extend the street car services, and have been adding and improving lines since. The streetcars have right-of-way at road intersections and roads have been narrowed to give more space to the light rail lines. This preferential traffic regulation has halved the travelling time for tram users compared to car users. The slogan of the tram company was 'faster than a sports car to the city centre'. Fares for the service are kept very low – for a set fare (about £22) people can enjoy unlimited travel on all public transport for a month.

There has been a positive response – the number of passengers has increased by over 100% since 1980. Trams now carry 59% of all passengers using public transport.

3 Improved roads

Some bundelungsstrassen (roads on which traffic can be concentrated) are needed, but are developed in a sustainable manner, e.g. Trunk Road 31, which had to be rerouted in the east of the city. This was done via two underpasses beneath the city.

4 Parking

Controlling parking is a key feature to the transport policy, so there is no free parking in the city centre. Commuters are encouraged to park-and-ride. In central residential districts there are resident-only parking zones, although even residents must pay for a parking pass. The evidence that this is working is clear – there are 4000 cars less per day entering the CBD than 30 years ago.

5 Bicycle use

In Freiburg the bicycle is essential for short journeys. The cycle-lane network has been extended from 29 km in 1970 to over 500 km today. There are over 5000 parking spaces for bicycles in the CBD, with more at each tram stop to encourage people to 'bike-and-ride'.

6 Compact settlement structure

Building density has been kept high, even in the suburbs to reduce sprawl, allowing the city to remain compact and thus possible to serve by the transit system. Building is

characterised by four- to five-storey apartment houses focused on the tramlines. A total of 70% of the population live within 500 m of a tram stop.

Have these measures helped ensure the sustainable development of Freiburg?

The investment and integration of land use and transport planning has led to a relative reduction in traffic congestion locally. The number of trips made by car in weekdays has stayed at the same level in comparison to 1976: the percentage of car trips relative to other modes of transport has dropped considerably.

The authorities are finding it increasingly difficult to completely relieve traffic congestion since regional transport policies are still automobile-orientated. However, the policies are helping this settlement develop a sustainable transport system, as many people are using public transport. The non-polluting trams allow today's commuters to travel quickly into the centre of Freiburg without damaging the chances of future generations to also use those routes.

web link & extra resources

www.eaue.de/winuwd/84.htm – a thorough write up on Freiburg's transport policy.

www.ourplanet.com/imgversn/121/bohme.html – provides an extension of this case study.

continues

activities

1 (a) State two reasons that account for increased traffic flow in urban areas (towns and cities). [2]

   (b) Explain why increasing numbers of cars on our roads is a problem. Include at least two reasons in your answer. [6]

2 Go to the website: www.freiburg-online.com and choose the English text version. Use this site to help you plan a day in the city. The only condition is that you must use a 'Regional Environmental Ticket'. [6]

3 By adding detail to each leg, complete this spider diagram summarising Freiburg's traffic management strategies. Add symbols to help your learning.

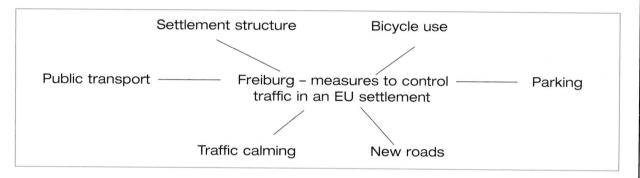

Settlement structure   Bicycle use

Public transport ——— Freiburg – measures to control ——— Parking
                        traffic in an EU settlement

Traffic calming   New roads

[12]

HIGHER TIER

4 Take on the role of an environmentalist. Write a paragraph evaluating the sustainability of Freiburg's transport policy. Include both sides of the argument in your answer. [8]

PLANNING FOR SUCCESS!

How do *you* learn?

Our brains process information in three different ways and this influences how we can best learn.

Figure 1
Types of learners.

 Some of us are VISUAL learners and learn better by looking at diagrams and pictures. We find it helpful to see things in colour and remember moving images and shapes.

 Others are AUDITORY learners who learn better by responding to voices and sounds.

 Many of us are KINESTHETIC learners who like to move when learning and find it helpful to touch objects or use computers when learning.

So how does this affect your revision?

We use all of these methods to learn material but if you think for a few minutes you may realise that one or two of these styles of learning seems more attractive to you. These are your *preferred learning styles* and you are likely to be more successful if you use your preferred style more often when you are learning work for your examination.

Top Tip

Think hard about what is the best way for you to learn and then use this approach more often to be successful.

The Learning Cycle

Geographers are very interested in processes, cycles and strategies. As you approach the examination it is important to have in place a strategy to help *you* prepare for the examination in an effective manner. Without a well-developed plan, and a reliable process by which you can revise, it would be too easy to become overwhelmed with all of the information and not do yourself justice in the examination. This is where the **learning cycle** comes in. There are four clear stages in learning material for the

Figure 2
Preferred learning styles.

Visual learners should:

- Write in different coloured pens and use highlighters;
- Use bullet points and key words;
- Use pictures, mind maps, diagrams, computers and flowcharts.

Auditory learners should:

- Tape notes and play them over;
- Discuss work with friends;
- Recite work aloud and make up raps and poems;
- Listen to music while learning.

Kinesthetic learners should:

- Use computers;
- If possible do things practically, feel and touch;
- Walk around while reading;
- Sequence your notes on pages or cards and order them to make sense.

examination and you must go through all four stages to improve your recall of facts.

Revision is an active process in which you must participate fully to achieve your potential and achieve the best examination result. Remember rereading is not revision. The following steps will help you remember all the facts you need to know to achieve a top grade.

Top Tip

Remember to use your preferred learning style when you are reworking and memorising material.

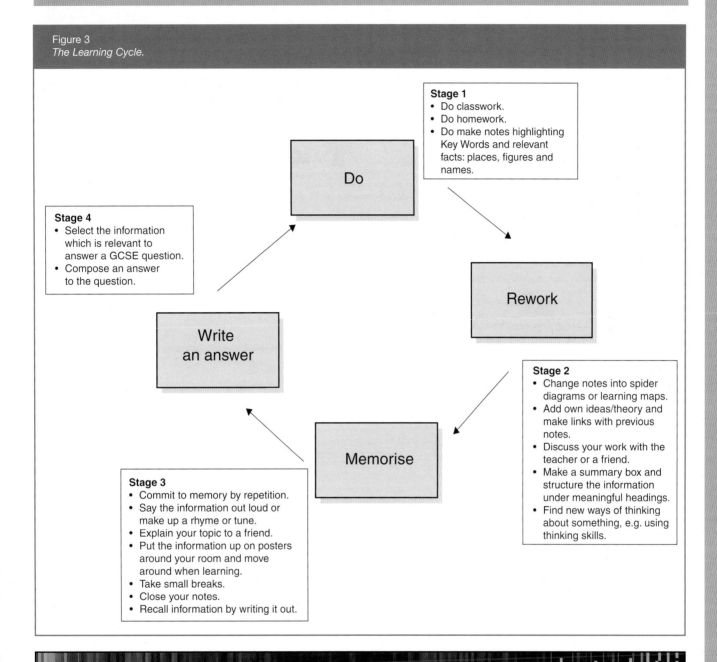

Figure 3
The Learning Cycle.

Stage 1
- Do classwork.
- Do homework.
- Do make notes highlighting Key Words and relevant facts: places, figures and names.

Do

Rework

Stage 2
- Change notes into spider diagrams or learning maps.
- Add own ideas/theory and make links with previous notes.
- Discuss your work with the teacher or a friend.
- Make a summary box and structure the information under meaningful headings.
- Find new ways of thinking about something, e.g. using thinking skills.

Memorise

Stage 3
- Commit to memory by repetition.
- Say the information out loud or make up a rhyme or tune.
- Explain your topic to a friend.
- Put the information up on posters around your room and move around when learning.
- Take small breaks.
- Close your notes.
- Recall information by writing it out.

Write an answer

Stage 4
- Select the information which is relevant to answer a GCSE question.
- Compose an answer to the question.

activities

1 Using the learning cycle, revise material on a topic to assist you in preparing for an examination question. For example, the following question taken from Theme F: Settlements and Change.

State fully two ways in which planners have attempted to create a sustainable urban environment in the inner city area of a city you have studied.

Top Tip

See the Case Study on regeneration in Belfast's Laganside area.

Reworking your notes into a more visual form

So we have all this information, but what do we do with it? Having gone through the learning cycle with all the relevant material, it is now time to reproduce your work in a more visual form to help you recall your work and to see the links both *within* each theme and also *across* themes. Spider diagrams are one way to help you visualise your work. Below is a spider diagram for Global Warming (Theme A, page 25). Learning maps or mind maps also help you to classify or organise information. This is a way of using key words which jog your memory on the topic.

Figure 5

Spider diagram map for global warming.

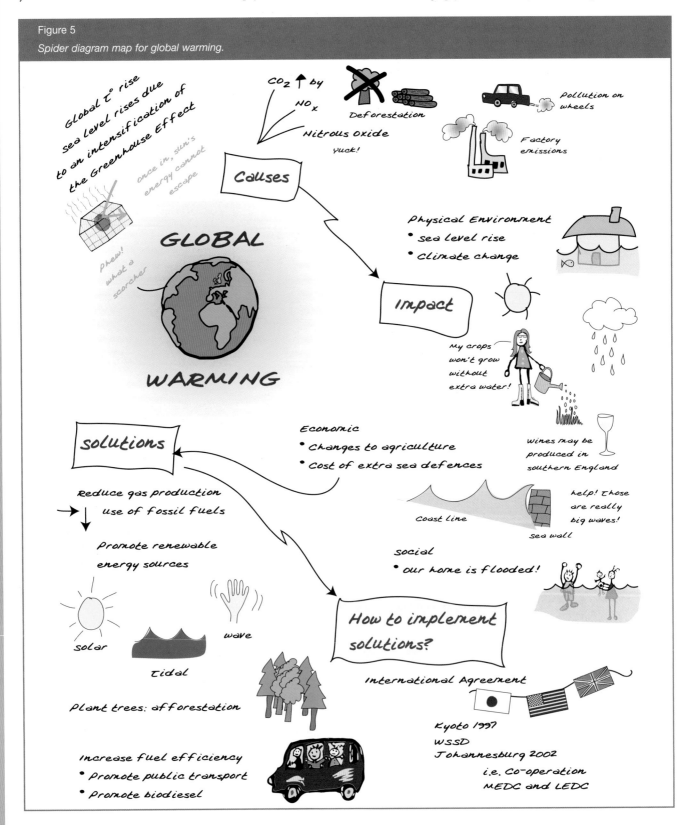

Learning maps use:

■ Lines to connect information;

■ Drawings/sketches/maps;

■ Grouped ideas.

Learning maps can be drawn at different scales and are also an effective way to memorise case studies. You can put each layer of information on the map in a different colour and increase the amount of detail the further you move away from the centre. All the hard facts – **fact figures**, **fact places** and **fact names** relevant to your case study, should be at the edge of the learning map. Overleaf is a learning map on our case study on page 37, an earthquake event in a MEDC – Kobe, Japan (Theme B).

Figure 5

Producing a learning map.

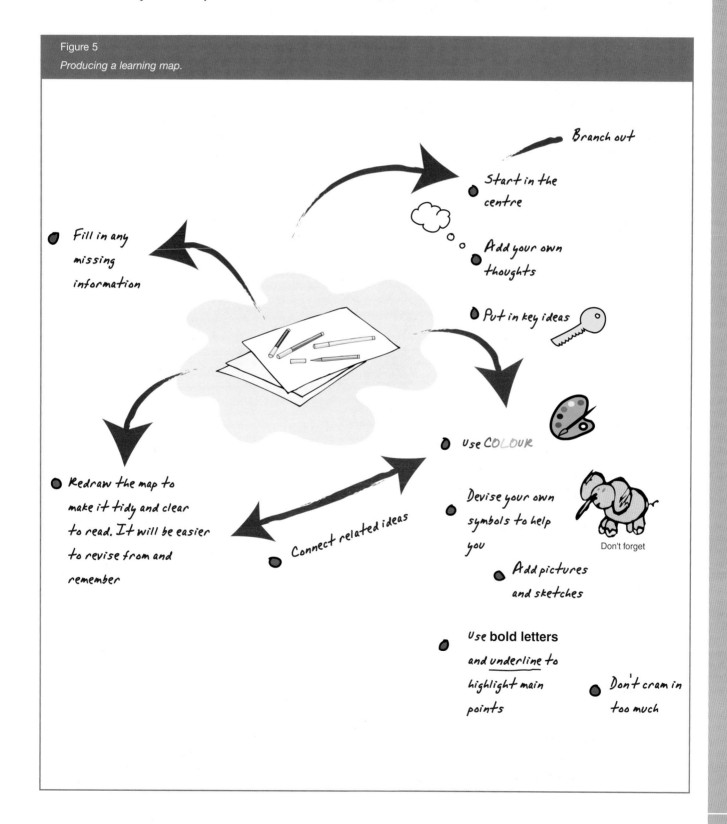

Figure 6

Learning map for Kobe earthquake.

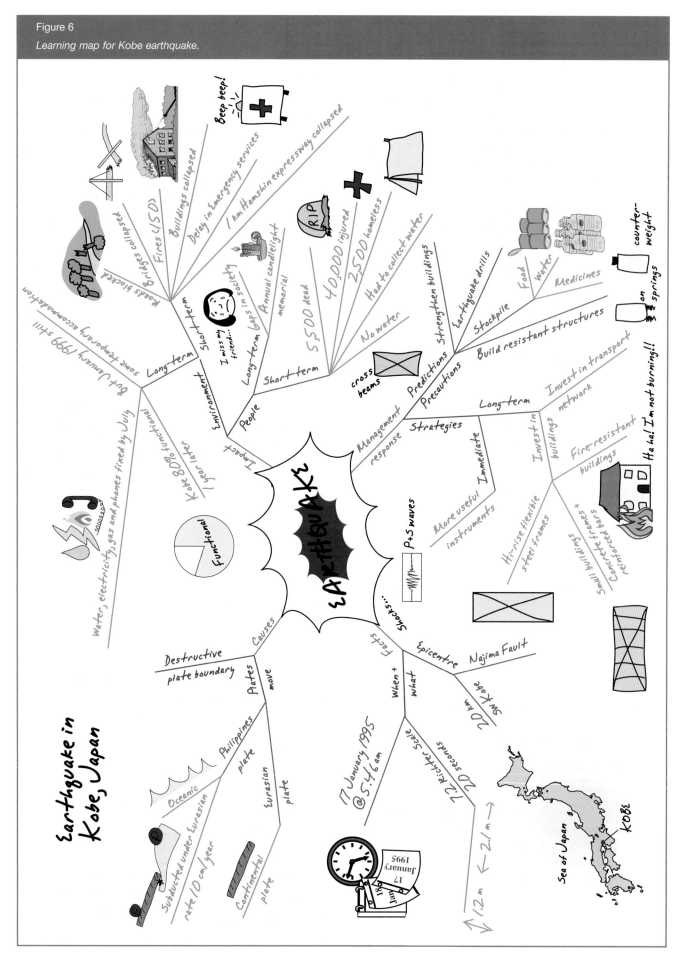

activities

1 Produce a learning map for your case study of an earthquake event in a LEDC, for example El Salvador (see page 39). The map should cover the same sort of information as that for the MEDC earthquake event in Kobe illustrated in Fig. 6. [10]

2 Produce a learning map for your case study of 'One renewable energy production scheme at a regional/national scale'. (See information on wind farms in Northern Ireland, Theme D, Unit Three, page 119.) [10]

Your learning map should contain information on the following:

■ What renewable energy is;

■ Location of your scheme;

■ What type of energy is produced;

■ How the scheme is a sustainable approach to energy production;

■ Benefits of the scheme for the environment and sustainable development;

■ Problems caused by the scheme to the environment or to sustainable development.

3 Learn your case study on wind farms in Northern Ireland using one of the strategies outlined for Stage 3 of the learning cycle. For example, explain it to a friend, or rewrite the information as bullet points on Post-Its then put them in order or repeat the information out loud.

Now try one of the following questions.

(a) State the location of a renewable energy production scheme you have studied. [1]

(b) State fully two benefits of this scheme for the environment. [6]

HIGHER TIER

4 More sustainable approaches to resource production must be found. With reference to a renewable energy production scheme at a regional/national scale that you have studied, evaluate the scheme in terms of sustainable development. [7]

What's in a word?

All of the questions on the exam paper will have one or more **command words** which tell you what the examiner expects you to do in answering the question. It is important that you read the question carefully to identify the command words, as otherwise you may not answer the question set. If you miss the point you will lose vital marks and if you write too much you are wasting precious time. Read the questions carefully and <u>underline</u> the command words to help you focus on what you are being asked to do and to ensure you score high marks!

On the next page there is a list of some command words commonly used in CCEA GCSE Geography examinations along with their meanings and an example of each.

activities

1 Look through last year's examination paper and for each question underline the command word and rewrite the question in your own words clarifying what it is asking.

Command Word	Meaning	Example
<u>Describe</u>	Write a descriptive answer on the topic without explaining it. You are often asked to describe differences, effects or consequences of something. If describe is used with a graph or table of data please remember to quote values in your answer.	1 **Describe** two social consequences of industrial change in a named inner city area you have studied. 2 Study the climate graph for Malta in the Mediterranean. *Source: The New Wider World*, David Waugh. Malta 36°N Altitude 18 m Annual range of temperature 13°C Annual precipitation 501 mm **Describe** the climate of Malta. Rainfall _____ Temperature _____
<u>Explain</u>	Develop your answer in the form of a reason for something or account for something.	**Explain** why a site close to a university can make an area attractive for a company wishing to set up a hi-tech industry.
<u>Describe and explain</u>	This requires a statement of fact plus an explanation of the reasons why.	**Describe and explain** the distribution of population in one country in the EU (outside the British Isles), which you have studied. (Tip – Spain was our Case Study.)
<u>Discuss</u>	Discuss requires you to examine or consider the facts and to make an evaluative comment. You will need to weigh up the advantages/ disadvantages or benefits/problems.	**Discuss** to what extent people per doctor is a good indicator of development.
<u>Label/Complete/ Match</u>	Used for completion of data response material. You may be asked to label diagrams, complete graphs or match statements in the form of tops and tails.	**Label** source, tributary, mouth, meander and floodplain on the diagram of a drainage basin.
<u>Name</u>	Identify something, for example a geographical feature/ term or place.	**Name** a producer and a consumer in this foodweb. (Data response.)
<u>State</u>	Requires you to write a short answer presenting a fact or facts without further explanation.	**State** the difference in the temperature between Glasgow and Moscow. (Data response.)
<u>State and explain</u>	Requires the statement of a geographical fact with a detailed reason of the same.	**State and explain** the main factor which causes the climates of Glasgow and Moscow to differ.
<u>State fully</u>	This question requires you to expand (or elaborate) on the aspect being examined. It is often used in relation to a reason for a geographical pattern or trend. (This term is used because you cannot explain a reason since a reason is in itself an explanation.)	Alaska is an area with a low density of population. **State fully** one reason why this is so.
<u>State the meaning</u>	Usually used for definitions. No further explanation is required other than showing the examiner that you know what the term means.	**State the meaning** of the term earthquake. This question merely requires a definition of an earthquake; not information on how or why earthquakes occur.
<u>Suggest</u>	This command word is often used when there may be more than one explanation and any relevant explanation offered would be acceptable.	1 **Suggest** an explanation for the cloud northwest of the British Isles. (Data response.) 2 Discharge varies along the main river channel. **Suggest** two reasons which might help to explain this.
<u>Give/Identify/ List</u>	These words are sometimes used in data response and knowledge questions. They expect you to examine data and extract information, or recall facts which you have learnt.	**List** two hard engineering measures taken on this river. (Data response.)
<u>Compare</u>	Requires comparative sentences on the items being examined, for example graphs, population pyramids or photographs. Be careful not to simply write two accounts or descriptions. Use comparison words such as *while, but* or *whereas* in your answer. Another example would be 'A is bigger but B is smaller'.	**Compare** the structure of the population pyramids for Portstewart and Ballymoney. (Data response.)
<u>Contrast</u>	Differences should be highlighted, e.g. differences in facts values or figures.	**Contrast** the difference in energy consumption between LEDC and LEDC regions. (Data response.)

Hitting the mark – what is the question really asking?

There are three main types of questions in GCSE Geography:

- The **Recall** Question, which is designed to test your knowledge.
- The **Stimulus/Data Response** Question, which is often linked to assessment of skills.
- The **Levels of Response** Question, which includes case study questions.

The examiner will also award marks based on the **quality** of your written answer.

Recall questions

These questions are designed to test your knowledge. To answer well you need to memorise knowledge and revisit it often so that you can recall it accurately and in detail. The learning cycle diagram can help in this:

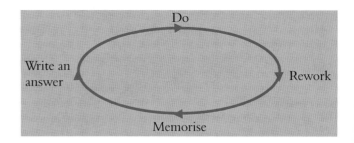

You must be able to recall specific facts and demonstrate locational knowledge about the places, environments, landscapes and themes you have studied. You may have to apply your knowledge to a different context or at a different scale. Having recalled the knowledge required you can then choose which facts are relevant to the question being asked.

Top Tip

Highlight words in the question to clarify what is being asked. You can also add your own words if it helps you to focus on relevant knowledge needed for the answer.

Example 1

Breakdown of rocks by chemical reactions (solution)

Give details on the solution process and its effects.

Question: **Chemical weathering** occurs in limestone areas. **Explain** how chemical weathering causes the **breakdown of limestone rock**.

How is it changed?

Example 2

Write about two causes of urbanisation (push and pull factors) which encourage people to move to the cites

Question: **Urbanisation** is taking place in many **LEDCs**. **State fully two** reasons why this is so.

increase in proportion of people living in towns and cities.

Less Economically Developed Countries.

The Data Response question

These questions may be based on an Ordnance Survey (OS) map, a photograph, a sketch map, a graph, a weather map, a cartoon, a newspaper article, a table of figures or some other type of geographical information. The questions set test your **observation** and **understanding** of the information given. Observation is what you see in the map or diagram, and understanding is what you conclude from your observation.

To demonstrate understanding you may have to apply your knowledge to describe and explain patterns or trends in the statistics, or draw conclusions from the information given. You may even have to compare two pieces of data. Sometimes stimulus response questions also test your skills, for example in the completion of a climate graph or flow diagram using data provided. Be careful to answer all such questions accurately. Many top candidates lose marks every year through careless mistakes on simple skills-related questions.

Information is often presented in tables or graphs. The following rules are useful when studying these diagrams.

Rules for tables and graphs

1 Read the stem of the question and the title of the diagram carefully to make sure that you understand the purpose of the table or graph.

2 Look at the column headings on tables or axes on graphs to be sure you know what they represent.

3 Think about the units of measurement on graph axes, e.g. cumecs in relation to discharge or percentages of total population in population pyramids.

4 Identify any patterns or trends, positive or negative, in the data. Pay attention to variation in the rate of change in a graph. Try to offer reasons for any anomalies – points on a graph/figure or in a table, which do not seem to fit in with the overall trend.

As you can see the answer you give to questions with tables or graphs really need you to **describe and explain** what the diagram shows.

Top Tip

Examine the stimulus material carefully as it will contain information required for the answer. If you are asked to read off data from tables or graphs be careful to be accurate to gain the marks.

Example 1: The following question appeared on a Higher Tier examination.

Study Fig. 7 below which shows population and energy consumption by continent and Fig. 8 which shows the world's fastest growing energy users 1986–96. Answer the questions which follow.

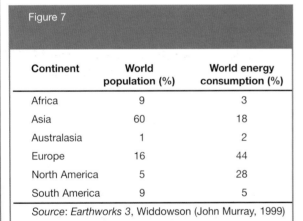

Figure 7

Continent	World population (%)	World energy consumption (%)
Africa	9	3
Asia	60	18
Australasia	1	2
Europe	16	44
North America	5	28
South America	9	5

Source: *Earthworks 3*, Widdowson (John Murray, 1999)

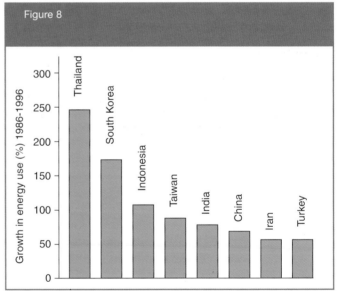

Figure 8

Remember to give values

Questions

(i) Using the table describe differences in the pattern of world population and world energy consumption. [4]

(ii) State fully the implication of the information on the graph for world energy use in the future. [3]

Following *Rule 1* for tables and graphs, you immediately see that the information in the table and that on the graph is on two different scales. The table gives us data relating to *continents* while the data on the graph refers to *countries*.

Moving on to *Rules 2 and 3* and on looking at the column headings, we see that the table gives information on two variables about each continent. We are told the population of each continent as a percentage of the total world population and also the percentage of the world's energy which that continent uses. This is important as it leads us to consider if there is any particular relationship between these two variables.

Similarly, by looking at the labels on the graph we see that it shows the percentage growth in energy use for the eight countries which have had the fastest increase in energy use between 1986 and1996.

Rule 4 leads us to attempt to identify patterns in the data. This is the basis from which we can attempt to answer the questions.

Question (i) tests your understanding of the table which having followed the rules is straightforward enough. The command word asks you to *describe differences* in the two variables. In other words, write about what the differences are in the pattern of world population and in the pattern of world energy consumption. Be careful – you are not asked to explain these differences. A good answer would be:

Continents such as Europe and North America which are composed of MEDC countries consume the largest percentages of world energy; e.g. North America is responsible for 28% of the world's energy consumption and Europe uses 44%. However this is not proportional to their populations as North America has only 5% of the world's

population and Europe 16%. In contrast, continents with LEDCs such as Asia has 60% of the world's population yet is responsible for only 18% of the world's energy consumption and Africa has 9% of the population and uses only 3% of the energy. There is clearly an imbalance between population size and energy consumption.

This answer is good because it draws attention to the differences (or compares) MEDC and LEDC continents in terms of both population and energy consumption, and it includes figures to justify each statement.

Question (ii) allows the examiner to test your ability to extract information from the graph and also apply your knowledge to develop your observations. Top students should be able to make a valid statement about the information on the graph and then write about a consequence of this, elaborating on its impact as in the following answer:

All of the countries listed have increased their energy consumption considerably between 1986 and 1996, e.g. in the case of Thailand the fastest growing in terms of energy use by as much as 250% and even over 50% in Turkey, ranked eighth on the graph. As these are mostly LEDCs which traditionally have used less energy, this growth will put even greater pressure on energy sources in the future and sustainable approaches to energy consumption must be found.

Example 2: A data response question from a Foundation Tier paper.

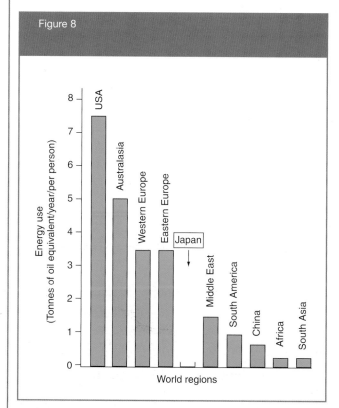

Study the graph left, which shows energy use per person around the world. Answer the questions which follow.

(i)   Complete the bar graph for Japan using information in the table below.   [2]

Region	Tonnes of oil equivalent/year/person
Japan	3 tonnes

(ii)   State the number of regions in which energy use per person is greater than 2 tonnes.   [1]

(iii)   Read the statement below and state whether it is true or false.   [1]

"The people in the Less Economically Developed countries in South Asia use more energy (oil) than do the people in the More Economically Developed countries of Western Europe."

Source: CCEA GCSE exam, Foundation Tier Paper 2, 2002.

Data response questions like these are designed to check if you understand the information presented on the graph and can use basic skills to complete the graph. Using *Rule 1* you should recognise that the graph is showing the amount of energy used by each person in different parts of the world. Applying *Rules 2 and 3* we discover, by looking at the axis of the graph, that the energy used is in tonnes of oil equivalent per person per year. This is because not everyone would use oil, but whatever types of energy they use in different places is changed into its equivalent in terms of oil so that comparisons can be easily made between people in different regions.

Think carefully about the information on the graph and you should see a pattern. Which countries, MEDCs or LEDCs, use the most energy? This will help you answer question (iii).

When answering the question be careful to be accurate. In part (i) you are being asked to complete a graph – a relatively straightforward task but one which a surprising number of candidates fail to answer, or they make silly mistakes. Always try to use a style very close to that already used on the graph. To achieve both marks in this question you need to draw the bar accurately on both axes. If you are more than 1 mm out you will be awarded only 1 mark. It is worth taking a little time and getting it right.

To gain full marks part (ii) also requires accuracy. The answer is 5, including the bar for Japan which you have just added. This is an example where care will ensure you achieve all the marks.

Having followed *Rule 4* you should have a fair idea of the pattern for question (iii). Just read the statement carefully to be sure you understand what it is saying. The answer is false because the MEDCs in Western Europe use much more energy than the LEDCs of South Asia.

Data response questions like these are often used by the examiner as an introduction to other questions, which will test your recall of knowledge and understanding. For example, this last question could be followed by a question on problems caused by increasing demands for energy in all regions of the world or how we can try to develop a more sustainable approach to energy use and how renewable energy sources can be developed.

Top Tip

Investigate to accumulate. Read the introduction to the question and any other titles. Look at column headings and labels on axes very carefully and decide if there are any clear patterns or odd ones out on your graph or table.

Ordnance Survey maps

There will be an Ordnance Survey question on one of the examination papers each year. The map will be an extract from a 1:50,000 OS map and while the symbols will be printed alongside the extract, it is obviously beneficial in terms of time if you are familiar with the symbols before you go into the examination. The examiner will be testing the following in OS questions:

- Your skills in map reading and interpretation.

- Your background knowledge and understanding of the theme which the OS map is presented with.

You may have to demonstrate skills in:

- Locating features using the grid lines and give four- and six-figure grid references;

- Using the scale to measure straight-line and curved distance between features;

- Giving compass directions, for example village X is south-west of town Y. Be aware of direction. Look carefully for the direction of North on the map. North is usually at the top of the map but it may not be in all cases;

- Recognising features from the symbols and contour patterns;

- Matching a photograph, overlay or sketch map to part of the OS extract.

In particular you will need to:

- Understand how relief is shown on the map through spot heights, triangulation pillars and contour patterns. This will enable you to interpret the shape of the land;

- Know what the Ordnance Survey 1:50,000 symbols mean, to identify things more quickly without constant reference to the key. This will save you valuable time in the examination.

Other questions will require you to write about (**describe**) patterns, for example the pattern of roads in an area. Look carefully and write what you see, for example the road pattern in or around a settlement or the fact that the road pattern follows low land. Routes (roads, railways or canals) should always be described in relation to other features, for example the settlements they connect or the relief of the land. Another example would be to describe the site of a settlement.

To **explain** something you need to give reasons for a pattern or activity which you have identified on the map. For example, you could be asked about the location of hi-tech companies in an area given the grid references of the same. Similarly, the reason why routes are often found on low land is because they are cheaper to build if they avoid having to climb steep hills or avoid land which is liable to flood. Railways were also cheaper to build where there was no need for cuttings or tunnels. To answer a question requiring explanation you will need to rely on what knowledge you have memorised and see if this knowledge helps you.

Occasionally you will have to complete a sketch map or cross-section. Make sure you line the sketch map up correctly with the map and complete it accurately.

The best way to prepare for the OS map question is to practise, practise, practise! Look at local maps and note the sizes on the map of settlements that you are familiar with, test your ability to identify features, and attempt to estimate distances and areas. In time you will become more confident using maps. They really are full of useful information and are quite easy to read when you have become familiar with their specific language.

For an example of an OS map and questions see Theme B, Glendun River study, page 43 and Theme E, Changes in function of industrial premises – Yorkgate, Belfast, page 128.

Photographs – a picture is worth a thousand words

Well not quite … but photographs are very important to geographers as they give an immediate picture of what a place is like and this means that long descriptions are not necessary. You may have used photographs in your coursework to illustrate your work and these are most effective when annotated to explain the geographical feature or characteristic which they show.

In an exam they may be used to test your:

- Ability to apply knowledge and understanding to a new situation, for example when used with an Ordnance Survey map. You could be asked to match up the photograph to a sketch map, model or cross-section. Overlays with the main features traced off onto tracing paper are often used to help you;

- Skill in identifying and describing something geographically, for example when labelling a diagram which corresponds to the photograph or when describing a geographical environment;

- Ability to analyse and interpret patterns. This means you need to be able to describe **what** geographical feature or pattern the photograph shows, **where** it is and then suggest an **explanation** for it. An example of this could be an aerial photograph of a settlement with a question asking you to suggest and explain some of its locational features;

- Understanding of geographical vocabulary, for example by describing the landscape or human activities, for example as evident in zones of a settlement, illustrated on the photograph.

> **Top Tip**
>
> Look carefully at the photograph and don't assume something is there if you cannot see it. Use only information which is relevant to the question and support your answer with map evidence.

Photograph of Station 1 on the Glendun River can be seen on page 45 and a photograph of the rural–urban fringe can be seen on page 188. See also the photograph on the location of Irwin's Bakery, page 125.

The weather map

When interpreting weather maps use the following checklist to explain different weather conditions when you are asked to explain the variations in the weather on a map.

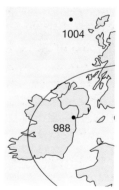

The weather system

- What pressure pattern is shown by the isobars? A depression with warm, cold or occluded fronts, or an anticyclone with high pressure?

- Consider what air mass is affecting an area. Is it warm or cold, Polar or Tropical? Is it wet or dry, Maritime or Continental?

Temperature

The temperature can vary with:

- Time of day;

- Season;

- Latitude or distance from the sea;

- The weather system, for example in a depression the distance from the fronts has a strong effect on temperature. Places within the warm front have higher temperatures than those behind the cold front. Places where skies are clear of cloud will be cooler at night due to heat loss by radiation.

- Winds, which affect temperature, for example southerly winds bring higher temperatures than northerly or easterly ones.

Wind speed

- This varies with the pressure gradient which is either gentle or steep. If the isobars are closely spaced then the pressure gradient is steep and the winds will be strong. The further apart the isobars are, the lower the wind speed.

Wind direction

- Winds blow clockwise and out of an anticyclone; anticlockwise and into a depression.

- Isobars bend at the fronts of a depression, indicating a change in wind direction at the fronts.

Cloud cover

- The amount of cloud in oktas increases where air is rising, for example along the fronts of a depression, or where air is damp, for example along coasts.

ICT

1 (a) Study the 3-day Europe 'surface pressure' animation for Europe found on the met office web site, www.meto.gov.uk.

   (b) State the name(s) of the weather system(s) affecting the British Isles. (2)

HIGHER TIER

   (c) Describe the weather you would associate with the main weather system on the chart. (Make sure you refer to temperature, wind speed and direction, cloud cover and precipitation in your answer.) (8)

   (d) Explain how and why the weather may have changed or remained stable over the 3-day period. (4)

- Anticyclones have less cloud cover. As the air in the anticyclone sinks, it warms up so it can hold more water vapour without allowing it to condense into water droplets.

The weather condition

- Symbols for precipitation must be explained by referring to the type of weather system.

- Rain is associated with rising air in depressions and is heavier at the steep cold front rather than the warm front.

- Fog and mist are associated with anticyclones, especially in the early morning where there has been very little cloud to trap the heat at night.

The Levels of Response Question

Levels of Response questions are the questions which generally require you to write at length. They are the questions which attract the most marks and include questions which are based on your case studies. These questions require you to recall information and to achieve top marks your answer must be specific to a particular place. Some levels of response questions require you to draw a sketch map, diagram or model. Obviously, more detail means more marks!

How would I recognise a levels of response question?

- Look at the marks. They are usually worth 6 marks or more.

- Look at the answer space. There are usually many lines left for your answer.

What does the examiner expect?

When the examiner marks a level of response answer he/she has to look for:

- Low Level Answers – answers which are correct but are simple and do not develop any points which have been made.

- Middle Level Answers – answers which are better because they have made more correct points or have taken at least one of the points and have developed it. Be careful. If the question asks for **one** reason, you need to develop your **one** reason. Do not provide an exhaustive list of reasons!

- High Level Answers – answers which have developed points in a clear and thorough way and show greater depth of understanding.

High Level Answers to questions such as 'State fully

how …', or 'State fully why …' should be tackled in the format:

- Make a **statement**

- Write about a reason or **consequence** of your statement

- **Elaborate** on your consequence or reason.

This is very important as it is exactly what the examiner will be looking for.

What is elaboration?

Elaboration can be in the form of further detail or an actual example, using facts on places, names or figures.

Get into the habit of making a statement then asking yourself 'So what?' Write a little more and then ask yourself 'So what?' again before writing another sentence or two. Try it. You will see it is a good help in elaborating your answers.

The following are some examples of questions which have levels of response answers. They should help you to see how to tackle these sorts of questions and get more marks.

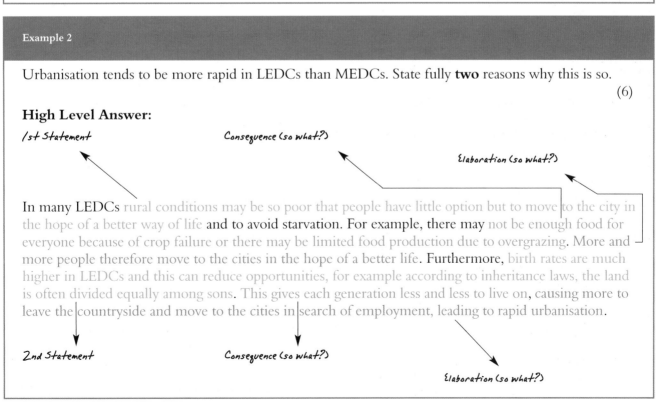

Example 1

State fully **two** problems that may occur in MEDCs as a result of immigration. (6)

A High Level Answer:

1st Statement made   Consequence (So what?)   Elaboration (So what?)

Immigrants who move into MEDCs are often willing to take lower paid jobs and this may mean native people are unable to get a fair wage for work in similar jobs which may make it impossible for them to maintain their standard of living. In addition, the immigrants may put a strain on social and health services in the country which can lead to racial tension between the newcomers and native people which can lead to crime and rioting which is bad for the economy.

2nd Statement made   Consequence (So what?)   Elaboration (So what?)

Example 2

Urbanisation tends to be more rapid in LEDCs than MEDCs. State fully **two** reasons why this is so. (6)

High Level Answer:

1st Statement   Consequence (so what?)   Elaboration (so what?)

In many LEDCs rural conditions may be so poor that people have little option but to move to the city in the hope of a better way of life and to avoid starvation. For example, there may not be enough food for everyone because of crop failure or there may be limited food production due to overgrazing. More and more people therefore move to the cities in the hope of a better life. Furthermore, birth rates are much higher in LEDCs and this can reduce opportunities, for example according to inheritance laws, the land is often divided equally among sons. This gives each generation less and less to live on, causing more to leave the countryside and move to the cities in search of employment, leading to rapid urbanisation.

2nd Statement   Consequence (so what?)   Elaboration (so what?)

A special word about case studies

The same rules apply for case studies but remember these answers must include elaboration on a spatial context. You need to show knowledge of real places. Your answer must be:

- Relevant to the question set;
- Have at least two **fact figures/fact names** or **fact places** relevant to the place you are writing about in your case study.

This is the main difference between middle level and high level answers. The high level answers have more specific locational facts. The facts are particular to the case study in your chosen area. For examples of case study answers see sections on the Foundation and Higher Tiers (pages 230–233).

How should I prepare for case study questions?

- Look over your detailed notes in your file/notebook.
- Highlight the facts – fact names, fact places and fact figures.
- Summarise information, for example in a learning map or rap.

Quite sure your Writing is Clear? – What is this QWC?

In examiners' speak QWC is:

Q – **Q**uality of

W – **W**ritten

C – **C**ommunication.

Basically this means 'How well can you express yourself' in answering the questions? QWC is assessed through the longer written answers which you give to questions on the examination paper, for example on levels of response questions and case study questions. Your aim should be to convince the examiner that you are a good geographer who can use specialist geographical vocabulary well in your answers. The specialist geographical terms will have precise meanings which save you time in the examination and show that you know your stuff! For instance it is much better and quicker to write 'transnational corporation' rather than 'a company which operates globally across national boundaries by having factories and offices in different countries', or 'discharge' instead of 'the volume of water in a river'.

To achieve high marks for QWC you need to be able to:

- Answer the question using relevant information;
- Write clearly so that the examiner can read your writing and know what you mean. Watch your spelling and punctuation;
- Use a suitable style of writing, for example try to write your answer fluently in a paragraph rather than using bullet points;
- Use a wide range of specialist geographical terms in your answers.

So how can I achieve this?

Be Prepared

Before the examination make sure you are have thoroughly learnt all the **Key Ideas** in each theme and can define them. You must also be familiar with all the **vocabulary** associated with each theme. You should know what all these terms mean and be able to use them. Make sure you can distinguish between terms such as LEDC and MEDC, urban and rural, physical and human, and economic, social and environmental.

In the Examination

Try to include the correct geographical term where possible in your answer and make sure the examiner knows what you mean.

Each theme has a set of **vocabulary** which you need to acquire and use in the examination. You will not be asked for definitions of these words unless they also appear in the list of **Key Ideas** for the theme, but good students should be able to use them correctly in their answers.

An example of a sentence which uses a high standard of QWC would be:

> The Transnational Corporation (TNC), Nike, has opened up in Taiwan, South Korea, Hong Kong, China, Indonesia, Vietnam and the Philippines and it therefore provides a good example of globalisation of industry.

This is better than attempting to explain TNC and globalisation of industry in your own words.

1   Read the vocabulary listed below which is associated with Theme E: Economic Change and Development.

Traditional industry, hi-tech industry, sunrise industries, information services, transnational corporation, infrastructure, investment, relocation, primary, secondary, tertiary, locational advantages, globalisation of industry, export.

2   Work through the learning cycle with these words.

(i)   **Do** highlight them in your revision notes.

(ii) **Rework** them by producing a simple revision glossary by adding your own theory words to explain the word or term.

(iii) **Memorise** them with your notebook closed.

(iv) **Write an answer** on your chosen theme using the correct vocabulary.

3   Repeat the previous two activities for each theme as you revise it.

AIMING HIGH IN YOUR EXAMINATION

To achieve a top grade in your examination follow these guidelines.

- Answer all of the questions. Even if you think you know little about the topic you can still pick up some marks. Don't be afraid to *state the obvious*!

- Fill all spaces. The examiner will have left you the right amount of space to answer the question. If you have lots of blank lines you probably have not written enough.

- Aim to score high level marks, for example 3/3 on levels of response questions. Remember make your point then ask yourself 'So what?', 'So what?'.

- When asked to describe something on a table or graph, always include values/figures from the table or graph in your answer.

- Include at least two facts about places in your Case Study answers. These can be place names, figures or other specific details.

- Know your key words well as the definition questions are a good way to pick up easy marks.

- Read the information given in data resource questions carefully as it often contains clues to the answer.

- Keep an eye on the time as you have to complete a variety of different types of question in the time given.

Aiming for A* on the Higher Tier

Here is an examination question. Read the comments around the question carefully. Try to write an answer for the question yourself before you read the sample answers. This will enable you to compare your answer to that of each of the two candidates.

Write about — Give reasons for — Consequences/results — Why would more money affect energy resources? — Need to write about places at the global or international scale

Describe and explain the impact of growing population and increased economic prosperity on world energy resources.

(6 marks)

A geographical term – need to know what it means.
A resource is a feature of the environment that is needed and used by people. Energy resources are those sources of power needed by people.

The combined effects of economic development as shown through economic prosperity, and population growth means that there is a growing need to manage the earth's energy resources.

Cathy's answer

The world's total population is growing with the population of LEDCs growing the fastest. This increase means that more energy is going to be needed in these areas.

Because many LEDCs are trying to change to be MEDCs they need more energy and therefore the current MEDCs are going to have to reduce their consumption.

At the moment MEDCs use more than 70% of the world's energy but they have only 25% of the population. Since no more energy can be created, MEDCs are going to have to reduce their consumption to make it more evenly spread.

Examiner's comments

Fair start – aware of population growth being high in LEDCs.

Makes some valid points but fails to write about their consequences and does not really elaborate on them. There is a hint that LEDCs are attempting to develop and in so doing will require more energy resources but this is not very well explained.

The reference to MEDCs having to reduce their consumption could relate to the fact that non-renewable resources will run out but this is not stated clearly.

No concept of renewable resources or alternatives in the answer.

Cathy makes no mention of particular countries at the global scale, just references to MEDCs and LEDCs.

Although the answer attempts to suggest what the impact will be – not enough energy for all – and quotes some fact figures, the lack of detail and development in the answer would mean that it would be marked middle level 2 answer.

Mark 3/6

QWC

Cathy uses few specialist terms in her answer. In fact she does not express herself very clearly, the examiner is left to work out what she means. Apart from this her spelling and punctuation are fine. If all of her answers were of this standard Cathy would get 3/6 for QWC.

Robert's answer

A growing population means there are more people needing energy resources. Increased economic prosperity also leads to further demands for energy to fuel the economy.

Most of the fuels used to produce energy in the world to-day are non-renewable or finite, e.g. oil and natural gas. This means that they will eventually run out. As population increases and more energy is used the world's energy resources are likely to become depleted unless they are managed carefully. This means that more sustainable methods of energy production such as wind, solar, and tidal need to be used. Where possible energy conservation should be encouraged in all countries.

Asia has 60% of the world's population, and uses 18% of the world's energy consumption. However many of its LEDC countries such as Thailand, Indonesia, Taiwan and South Korea are experiencing rapid economic growth and are becoming NICs (newly industrialised countries). This has increased the amount of energy they have used by >100% in 10 years and has put increased demands on energy resources. This demand will continue to increase because most LEDCs have high birth rates and more people means further pressure on energy resources.

In MEDCs, such as the UK and USA, industry continues to use large amounts of energy and as people become wealthier they purchase more domestic appliances and many homes have 2 cars, which also puts pressure on energy resources. Although USA has only 5% of the world population it uses a massive 28% of total world consumption because of its economic prosperity.

Examiner's comments

Shows candidate understands the focus of the question.

Good answer – Robert identifies an impact and links it to population growth.

Robert makes a valid **statement** here 'most … are non-renewable, e.g. oil, etc', and then gives a **consequence** of this "eventually run out … unless managed carefully".

He then **elaborates** on this by giving examples and referring to sustainable methods of energy production.

Specific case study facts at global scale, has more than two fact figures, fact places or fact names.

Good use of specialist terms throughout answer, e.g. non-renewable, finite, depleted, sustainable, NICs and birth rates.

Marks 6/6

QWC

Robert has used a variety of specialist terms correctly in his answer. He has also no spelling mistakes and has good punctuation throughout. His answer includes relevant information and is fluently written in prose and if the remainder of his answers were of a similar quality he would achieve 6/6 for QWC.

Aiming high on the Foundation Tier

The following is a typical question from a foundation tier examination paper. Read the information given about the question and try to write an answer for the question before you look at the answers given. This will help you to decide how good your answer is.

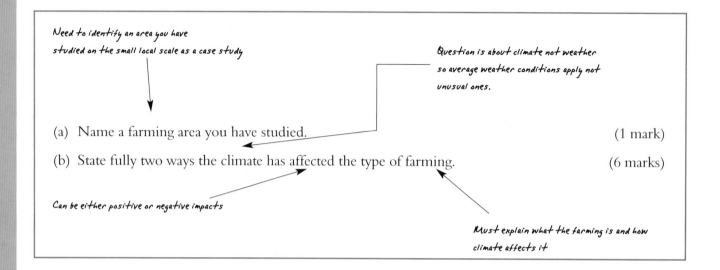

Need to identify an area you have studied on the small local scale as a case study

Question is about climate not weather so average weather conditions apply not unusual ones.

(a) Name a farming area you have studied. (1 mark)

(b) State fully two ways the climate has affected the type of farming. (6 marks)

Can be either positive or negative impacts

Must explain what the farming is and how climate affects it

Jack's answer

(a) Hill farm in Co. Antrim.

(b) Sheep farming is carried out because of the heavy rain. Animals are kept indoors in Winter because of the cold. The cold climate shortens the growing season.

QWC

This is a mid level answer in terms of QWC. There are no spelling mistakes but limited use of specialist terms.

Examiner's comments

(a) 1 mark awarded for area.

(b) The remainder of the answer is very general. There is no mention of figures, e.g. temperatures or amount of rainfall.

Still, Jack has linked the climate to the type of farming and has made valid statements. To gain more marks he would need to elaborate by giving more specific detail.

1 mark for heavy rain and 2 marks for impact of cold winter. Jack could have gained more marks by linking the cold temperatures (quoting figures) to the growing season and explaining its impact on the food source for the sheep.

Marks 3/6

This answer scores more marks because it is more specific.

Alicia's answer

(a) Beef Farm at Florencecourt, Co. Fermanagh.

(b) The beef cattle on this farm are kept indoors from mid October to mid May because the temperatures are not high enough for grass to grow. For example average temperatures drop from 10°C in October to 4.1°C in January. The growing season is only 240 days,

The area also receives heavy rainfall in Winter with many months having over 90 mm of precipitation. This means the cattle have to be kept indoors because if they were kept outside their hooves would plough up the ground making the fields unusable for some time.

Examiner's comments

(a) 1 mark – specific details.

(b) A good start clearly identifies the type of farming and links an aspect of it to the average temperatures.

Supports answer with fact figures.

When giving the second affect Alicia clearly states the link between rainfall and the housing of the cattle, she gives a reason for this and elaborates on the impact. Again she uses figures to illustrate.

Mark 6/6

QWC

Alicia writes fluently and uses appropriate terms and punctuation. Her spelling is also good. If the remainder of her paper was written in this style and included specialist terms she would achieve high level marks for QWC.

GO FOR IT ... Seven Top Tips

Get ready before the examination

- Know which themes are tested on each paper – Physical Themes (Themes A, B and C) on Paper 1 and Human Themes (Themes D, E and F) on Paper 2.

- Check that you have all the relevant notes.

- Catch up on notes you may have missed and make sure you understand them.

- Use the Learning Cycle to help you revise and produce your learning maps – these will be invaluable on the night before the examination.

- Look over past examination questions and locate each topic or generalisation tested on the paper.

On the day of the examination

- Explore the question.

- Identify the command word.

- Think about what information is requested and be selective about what you include. Don't simply write all you know about the topic. Answer the question set.

- Use the space wisely. There are 2 lines per mark for answers on the higher tier papers and 1½ lines per mark on the foundation tier. You will not be penalised for writing more but it is generally not necessary and you could be wasting vital time!

- Look carefully at the question to see how many responses you have to make. For example is it **one** reason or **two** reasons?

Formulate your answer

- Write down prompt words or clues to help you.

- Keep thinking about the question.

Organise the detail required for your answer.

- Elaborate on your initial statements – SO WHAT? …. SO MORE MARKS!!

- Remember: Statement – Consequence – Elaboration.

Read your work carefully

- Have you included all the information and answered the question?

Include all the relevant Fs to support your answer:

- Fact names;

- Fact places;

- Fact figures;

- Fact dates.

Timing is crucial

- Watch the clock to ensure you have enough time left to answer all questions.

- Remember half an hour per theme or full question!

Coursework issues

As part of your GCSE Geography examination you will be required to submit a piece of coursework based on fieldwork you have carried out. Although you may have worked as part of a group when collecting your data, you have to write up your report individually.

Fieldwork is a great way to discover more about geography and as a bonus the coursework itself is worth 20% of the marks for the course. Your coursework report should be less than 2500 words, so don't be tempted to overdo it! Nevertheless, you should make sure that you have included mention of all the points in the coursework checklist.

Coursework checklist

Make sure you include:

- Planning – this should mention the objectives of your study and identify the information you require;
- A title page;
- A contents page;
- Introduction – clear aims, a description of the methods you used and a description of the spatial context of your study. Include a location map;
- Aims and hypotheses;
- Maps of your study area;
- Methods of data collection and the equipment you used;
- Diagrams and graphs all with a title and carefully labelled;
- Relevant photographs – make sure you label them and explain what they are showing;
- Analysis for each aim – this is a description of **what** you found;
- Interpretation for each aim – an explanation of **why** you found what you did. Try to give reasons for the trends in your results;
- Conclusions for each aim;
- Overall conclusion – refer back to your aims and summarise your findings;
- Evaluation including the strengths and weaknesses of the study;
- Bibliography;
- Appendices – these should contain your data collection sheets and planning.

Presenting your data

Try to include a variety of geographical techniques in the data presentation section of your report to achieve high marks. Suitable techniques include:

- Scattergraphs and best-fit lines;
- Line graphs;
- Bar graphs and histograms;
- Pie graphs;
- Proportional circles;
- Flow line diagrams/maps;
- Choropleth maps;
- Isopleth maps;
- Rose diagrams;
- Sketch maps.

ICT is useful for producing a professional looking report, but remember if you use ICT to present your data you need to produce your graphs individually.

Do try to produce your best work as it adds greatly to your overall score. Fieldwork is one of the most enjoyable aspects of geography and it is an excellent way to revise part of the course.

Have you thought of a career using geography?

Having completed your GCSE course you will now be planning for your future and thinking about possible career options. Geography is a useful subject at GCSE and can lead to a wide variety of careers. While studying geography you will have acquired a range of desirable skills including:

- Good oral and written communication skills.
- Ability to analyse data and extract trends and patterns.
- Ability to explain findings and draw conclusions.
- Improved problem solving and decision making.
- Computer literacy.
- Sound literacy and numeracy skills.
- Skills in working as part of a team.
- A spatial awareness which is complemented by a social and environmental awareness.

This means that you have many job opportunities open to you. Some jobs are directly linked to geography while others make use of your skills and knowledge.

Jobs directly linked to geography

Some geographers are teachers or lecturers while others choose to work with GIS (Geographical Information Systems), or in environmental or development projects.

Jobs which use your geographical skills and knowledge

By studying geography at a higher level you will get the opportunity to specialise in an area of the subject which particularly interests you and this can open up many career options.

■ **Mapwork and Computing**

Geographers work with Geographic Information Systems (GIS) as cartographers and surveyors.

■ **Development**

Some geographers are employed by Aid agencies, government and non-government organisations, e.g. VSO.

■ **Population**

Geographers can be demographers, market researchers, census officers and social workers.

■ **Ecosystems**

There are opportunities in forestry and environmental consultancy.

■ **Settlement and Industry**

Geographers are employed as town planners, chartered surveyors and transport planners.

■ **Tourism**

Jobs are found in travel agencies, or tourist offices.

■ **Land and Water Processes**

Geographers also work as hydrologists, geomorphologists and civil engineers.

Further information can be found on the website of the Royal Geographical Society at www.rgs.org

Queen's University Belfast

School of Geography

© School of Geography

Training for the future:
Acquiring skills through Geography

While studying Geography, students attain knowledge and understanding of many elements of the subject, including environments and environmental change, settlement and urbanisation; and through this develop and strengthen their skills base.

Discipline-specific skills

Plan and execute research & study

Social survey & interpretive methods

Prepare maps and diagrams

Effective fieldwork

Technical & laboratory techniques

Debate ethical & moral issues

Intellectual skills

RESULTS

ORIGINAL DATA

REFERENCE MATERIAL

1964	9	3	0
1970	29	13	2
1975	65	33	10
1980	138	69	25
1985	296	114	59
1990	437	138	78
1995	520	166	97
1998		193	109

Developing reasoned argument and independent thought

Decision making

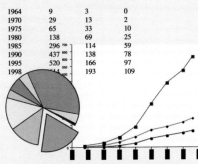

Critical evaluation & interpretation of data

Abstracting & synthesising information

Analysing and problem solving

Key and personal skills

Learning and study

Spatial awareness and observation

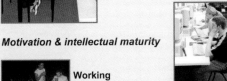

Information technology

Empathy, insight & integrity

Motivation & intellectual maturity

Self-awareness & self-management

Flexibility, adaptability and creativity

Working as part of a team

Numeracy & computation

Written and oral communication

KEY TO CASE STUDIES

Some of these case studies listed have already been provided in a case study format for you. Information necessary for all case studies is contained in this book as indicated below.

THEME A

A1	BRITISH ISLES WEATHER	PAGE 2
A2	CLIMATE OF CONTINENTAL EUROPE★	PAGE 17
A3	THE EXAMPLE OF EAST ANGLIA IN THE UK	PAGE 21
A4	THE EXAMPLE OF FARMING IN THE MEDITERRANEAN	PAGE 23
A5	GLOBAL WARMING, WORLD SCALE★	PAGE 25

THEME B

B1	GLOBAL PLATE MOVEMENTS★	PAGE 30
B2	AN EARTHQUAKE EVENT IN A MEDC – KOBE, JAPAN	PAGE 37
B3	AN EARTHQUAKE EVENT IN A LEDC – EL SALVADOR, CENTRAL AMERICA	PAGE 39
B4	GLENDUN RIVER, NORTHERN IRELAND	PAGE 43
B5	A RIVER MANAGEMENT SCHEME AT THE NATIONAL LEVEL – THE MISSISSIPPI RIVER, USA	PAGE 52
B6	A LIMESTONE ENVIRONMENT, THE BURREN, CO. CLARE, IRELAND	PAGE 59

THEME C

C1	A SMALL-SCALE ECOSYSTEM, BELVOIR PARK, NORTHERN IRELAND	PAGE 72
C2	TROPICAL RAINFOREST – THE AMAZON RAINFOREST IN BRAZIL	PAGE 74
C3	THE AMAZON RAINFOREST	PAGE 77
C4	A PEATLAND ECOSYSTEM, CUILAGH MOUNTAIN, NORTHERN IRELAND	PAGE 80
C5	ECOTOURISM IN KENYA	PAGE 84

THEME D

D1	GLOBAL DISTRIBUTION AND DENSITY OF POPULATION★	PAGE 90
D2	DISTRIBUTION AND DENSITY OF POPULATION IN SPAIN	PAGE 93
D3	GLOBAL POPULATION CHANGE★	PAGE 94
D4	POPULATION CHANGE AND STRUCTURE IN THE MEDC ITALY	PAGE 107
D5	POPULATION CHANGE AND STRUCTURE IN THE LEDC MEXICO	PAGE 109
D6	POPULATION RESOURCE RELATIONSHIPS AT THE GLOBAL SCALE★	PAGE 113
D7	A RENEWABLE ENERGY PRODUCTION SCHEME, WIND POWER IN NORTHERN IRELAND	PAGE 119

THEME E

E1	CHANGE IN FUNCTION OF INDUSTRIAL PREMISES, YORKGATE, BELFAST	PAGE 128
E2	HI-TECH INDUSTRY WITHIN THE BRITISH ISLES	PAGE 133
E3	A TNC AT THE GLOBAL SCALE: NIKE★	PAGE 144
E4	A SUSTAINABLE DEVELOPMENT PROJECT: FISHING IN SOUTH-WEST INDIA	PAGE 153
E5	GLOBAL TRADE AND AID★	PAGE 156

THEME F

F1	SETTLEMENT SITE FACTORS★	PAGE 170
F2	URBAN GROWTH AND CHANGE AT THE GLOBAL SCALE★	PAGE 179
F3	URBAN SETTLEMENT IN A MEDC, BELFAST, NORTHERN IRELAND	PAGE 189
F4	URBAN SETTLEMENT IN A MEDC, SÃO PAULO, BRAZIL	PAGE 195
F5	AN URBAN PLANNING INITIATIVE, LAGANSIDE, BELFAST	PAGE 204
F6	SOLUTIONS TO TRAFFIC PROBLEMS IN FREIBURG, GERMANY	PAGE 211

★Global case study and therefore not shown on the world map.

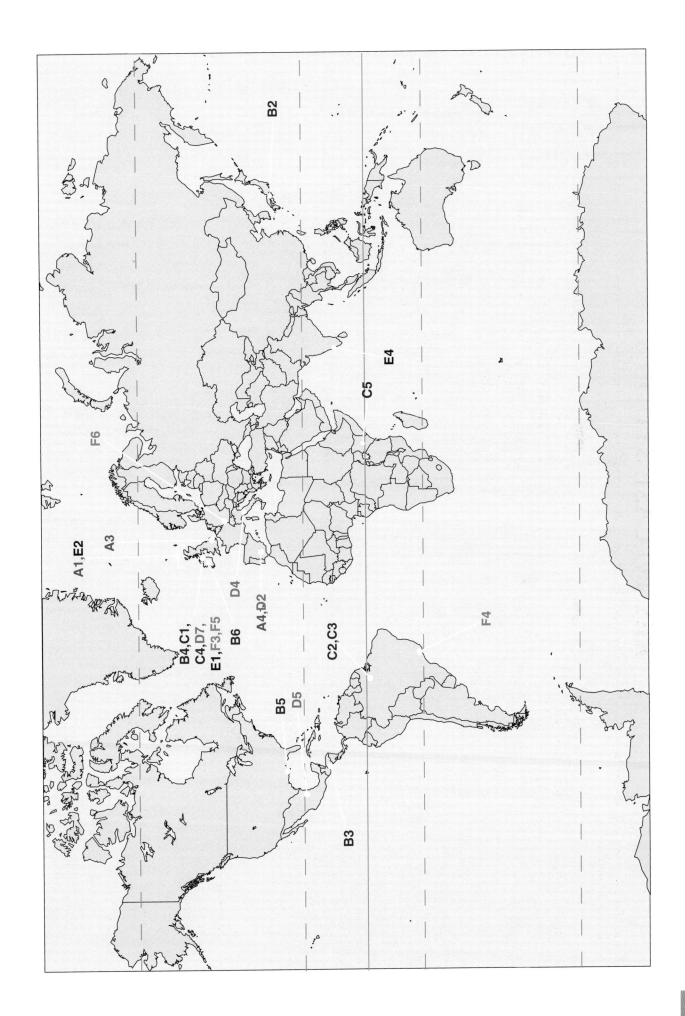

A1,E2

A3

F6

B4,C1,
C4,D7,
E1,F3,F5

B6

D4

A4,D2

F4

B5

D5

C2,C3

B3

B2

E4

C5

239

ORDNANCE SURVEY MAP SYMBOLS

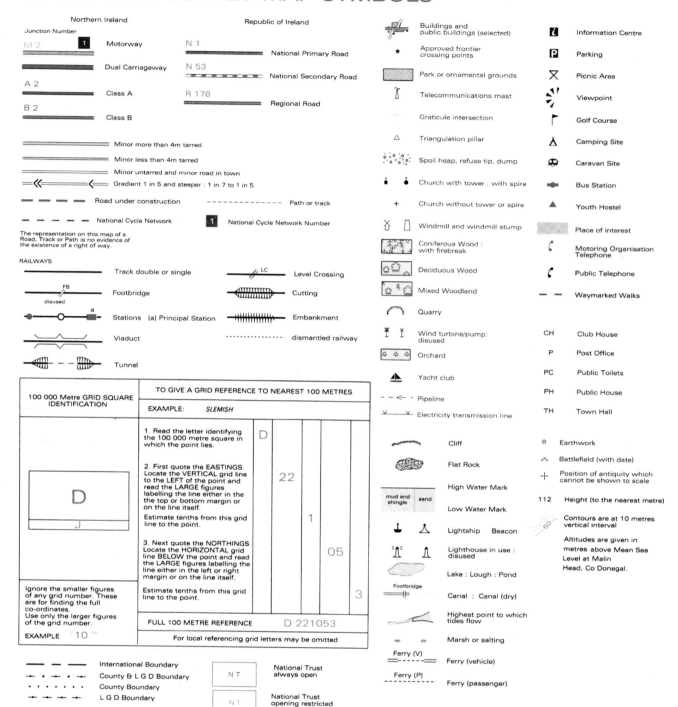

Northern Ireland

Junction Number

M 2 ┃1┃ Motorway

Dual Carriageway

A 2 Class A

B 2 Class B

Minor more than 4m tarred

Minor less than 4m tarred

Minor untarred and minor road in town

Gradient 1 in 5 and steeper : 1 in 7 to 1 in 5

Road under construction

National Cycle Network

The representation on this map of a
Road, Track or Path is no evidence of
the existence of a right of way.

Republic of Ireland

N 1 National Primary Road

N 53 National Secondary Road

R 178 Regional Road

Path or track

┃1┃ National Cycle Network Number

RAILWAYS

Track double or single

FB Footbridge

disused

Stations (a) Principal Station

Viaduct

Tunnel

LC Level Crossing

Cutting

Embankment

dismantled railway

100 000 Metre GRID SQUARE IDENTIFICATION	TO GIVE A GRID REFERENCE TO NEAREST 100 METRES
	EXAMPLE: *SLEMISH*
D	1. Read the letter identifying the 100 000 metre square in which the point lies. D
	2. First quote the EASTINGS. Locate the VERTICAL grid line to the LEFT of the point and read the LARGE figures labelling the line either in the the top or bottom margin or on the line itself. 22
	Estimate tenths from this grid line to the point. 1
	3. Next quote the NORTHINGS. Locate the HORIZONTAL grid line BELOW the point and read the LARGE figures labelling the line either in the left or right margin or on the line itself. 05
Ignore the smaller figures of any grid number. These are for finding the full co-ordinates. Use only the larger figures of the grid number.	Estimate tenths from this grid line to the point. 3
EXAMPLE ¹10	FULL 100 METRE REFERENCE D 221053
	For local referencing grid letters may be omitted

International Boundary

County & L G D Boundary

County Boundary

L G D Boundary

N T National Trust always open

N T National Trust opening restricted

Buildings and
public buildings (selected)

★ Approved frontier
crossing points

Park or ornamental grounds

Telecommunications mast

Graticule intersection

△ Triangulation pillar

Spoil heap, refuse tip, dump

Church with tower : with spire

+ Church without tower or spire

Windmill and windmill stump

Coniferous Wood :
with firebreak

Deciduous Wood

Mixed Woodland

Quarry

Wind turbine/pump:
disused

Orchard

Yacht club

Pipeline

Electricity transmission line

Cliff

Flat Rock

High Water Mark

mud and shingle | sand

Low Water Mark

Lightship   Beacon

Lighthouse in use :
disused

Lake : Lough : Pond

Footbridge

Canal : Canal (dry)

Highest point to which
tides flow

Marsh or salting

Ferry (V)
Ferry (vehicle)

Ferry (P)
Ferry (passenger)

ℹ Information Centre

P Parking

✕ Picnic Area

Viewpoint

Golf Course

Camping Site

Caravan Site

Bus Station

▲ Youth Hostel

Place of interest

(Motoring Organisation
Telephone

(Public Telephone

Waymarked Walks

CH Club House

P Post Office

PC Public Toilets

PH Public House

TH Town Hall

☆ Earthwork

Battlefield (with date)

Position of antiquity which
cannot be shown to scale

112 Height (to the nearest metre)

Contours are at 10 metres
vertical interval

Altitudes are given in
metres above Mean Sea
Level at Malin
Head, Co Donegal.

240

Afforestation 82
Age dependency 105, 107
Aid 163
Air masses 7
Air temperature 3, 5
Albania, out-migration 103
Amazon rainforest 74, 77
    conservation 78
Anemometer 2
Animals, in an ecosystem 73, 75,
    81
Anticyclones 11, 15
Appropriate technology 152
Asylum seekers 111
Atmosphere 17, 25, 26

Barograph 2
Barometer 2
Bed load 45
Belfast
    functional land-use zones 186,
        189
    premises change of use 128
    regeneration project 204
Bio diesel 28
Biodiversity 75, see also Ecosystem
Biomass 67, 71
Biomes 66
Birds, in an ecosystem 81
Birth rates 94
Brazil
    rainforests 74, 77
    São Paulo, urban settlement
        195
Burgess model 183

Carbon dioxide 26
Central business district 183, 186,
        189, 193, 199, 206, see also
        Functional zones
Choropleth maps 91
Cities
    skyline 186
    world's biggest 180
Climate 2
Climatic factors
    altitude 17
    continentality
    effect on farming 21, 23, 24
    European 18
    graphs 20
    in an ecosystem 67, 72, 74, 81
    latitude 17
    prevailing winds 18
    use of technology 24
Clouds 5

Coffee, use of fair trade 162
Comic Relief 164
Command words 219
Commodity trade 142
Communities, impact of
        development 206
Commuters 212
Conservation 65, 78, 83
Consumers 69
Consumption 115
Continentality 17
Convection currents, see Plate
        movement
Counter urbanisation 181
Coursework 235

Decomposers 69
Death rates 94
Deforestation 77, 162
Degrees Celsius 2
Demographic Transition Model 98,
        99, 101, 102
Depressions 5, 8, 9, 10
Desert 67
Development 147, 151
Development Diamond 150
Drainage basin 42, 47

Earth, structure 30
Earthquakes 31, 34
    effects of 37, 39
    epicentre 36
    focus 36
Ecology, see Ecosystems
Economic activities, types of 122
Economic factors 91
Economic indicators 147
Ecosystem 66
    examples in Northern Ireland 72,
        80, 83
    Brazilian rainforest 74
    see also Ecotourism
Ecotourism 85
Emigration 103, 110
Employment 193, 197, 206
Energy
    exploitation and consumption
        117
    renewable energy 119
Environment, threats to 62
Ethnic areas, see Socio-
        economic/ethnic areas
Evapotranspiration 42
Examination
    guidelines 229
    questions 221

Export, see Trade

Fair trade 159, 162
Farming, effect on the environment
        63
Field sketches 46
Flora and fauna 66, 69, 72, 74, 80,
        83
Flooding 50, 52
    control 51
    defences 55
    impact of 51
Fluvial processes
    deposition 46
    erosion 46
    transportation 46

Food chain 69
Food web 70
Fossil fuels 117
    effect on envirnoment 26
    see also Energy, Global warming,
        Transportation
Fronts 5, 8, 10
Functional zones, see Land-use zones

Gender-related Development Index
        150
Geography as a career 235
Global warming 25
    causes of 26
    effects of 28
Globalisation 138
Green belts 187, 198
Greenhouse effect 26
Gross national product (GNP) 142,
        147

High pressure systems, see
        Anticyclones
Hi-tech Industry 133
    location in British Isles 135
Housing 172, 180, 183, 184, 185,
        192, 193, 197, 199, 206
Human Development Index (HDI)
        147
Human impact 77, 81
Human Poverty Index (HPI)
        150
Human pressure, effects on the
        environment 62
Human welfare 147
Hydrological cycle 42

Igneous rocks 56
Immigrants 103, 110

Industry
location 125, 133
premises, change in function 128
primary 122
quaternary 122, 134
relocation 128, 141
secondary 122, 134
tertiary 122
traditional 125, 153
Infiltration 43
Information services, *see* Hi-tech
industries, Industry, quaternary
Infrastructure 130
Inner-city area 125
Insects, in an ecosystem 75, 81
Interdependence 156, *see also*
Globalisation
Internet use 140
Investment 133, 135, 138, 141
Ireland, limestone areas 59
Irrigation 23, 24
Isobars 3, 4, 8
Italy, population change 107

Japan, earthquake experience 37

Kenya, assisting through aid 165
Kenya, National Parks 86, 88
Kenya, tourism in 85
Kyoto Agreement of 1997 29

Land values 186
Landfill tax 209
Land-use/functional zones 183, 185,
186, 189, 196
Leaching 71, 72
Learning Cycle 214, 215
Learning maps 217, 218
Learning styles 214
Less economically developed country
(LEDC) 29, 101, 105, 107, 115,
141, 143, 147, 151, 158, 159,
162, 180, 181
city changes 200
city characteristics 192
earthquake response 41
exports 157
Life expectancy 99, 101, 107
Lignite 117
Limestone processes/features
caves and caverns 60
clint 60
features 59
gryke 60
pavement 60
stalactites 61
stalagmites 61
structure 58

swallow holes 59
threats 62
Litter 71
Low pressure systems, *see*
Depressions

Maasai tribesmen 86
Management of ecosystems, *see*
Ecosystem, Ecotourism
Meanders 48
Mediterranean biome 67
Mediterranean farming 23
Mega-city 182
Metamorphic rocks 57
Mexico, population change 109
Migration 103, 110, 181
Million (Millionaire) cities 180
Mississippi River, flooding 52
More economically developed
country (MEDC) 29, 101, 105,
107, 115, 141, 143, 147, 162,
180, 187
changes in cities 199
earthquake response 41
trade patterns 156
Multicultural society 112

Natural increase 94
Northern Ireland
example of an ecosystem 72
settlement hierarchy 174
use of fossil fuel 117
wind farms 119
see also Belfast
Nutrient cycle 71
Nutrients 69, 76

Ocean trench, *see* Plate boundary
Ordnance Survey maps 44, 177, 129,
221, 224
Overpopulation 95, 113

Pacific Ocean, volcanic activity 31
Peatland ecosystem 80
Photographs 46, 225
Physical factors 91
Planning 202
Plate boundary 30, 32, 37
Plate movement 30
Plate tectonics theory 30
Population 182, 195
aged dependent 107
change 94, 103, 107
Change Model 98
composition (structure) 99
density 90, 92
distribution 90
growth 95

independent 105
pyramids 101, 102, 103, 104, 105,
107, 109
working 105
Precipitation 2, 5, 42
Pressure 2, 3, 4
patterns 5
Primary producers 69
Producers, *see* Primary producers
Project work 46

Rain Gauge 2
Rainfall 5, 19, 23
cyclonic 5
relief 5, 6
Rainforest 66, 74, 77
cross-section 75
Range 174
Recycling 208
of nutrients 69, 71
Refugees 111
Resource
depletion, *see* Energy
non-renewable 116
renewable 119
Reuse 208
Richter scale 36
Rift valleys 33
Ring of Fire 31
Rivers 43
courses 43, 45
deltas 49, 54
depth 45
discharge 45
estuary 49
floodplain 49
levées 49
load 45
management 51, 52
mouth 45, 49, 54
source 41, 45
tributary 42
underground 60
watershed 42
width 43
Rock type
igneous 56
metamorphic 57
sedimentary 56
Rural–urban fringe 187, 188
Rural–urban migration 181, 182

Safaris 87, 89
São Paulo, urban settlement 195
Satellite images 12, 15
Savanna 66
Sector model 184
Sedimentary rocks 56

Seggregation, *see* Socio-
economic/ethnic areas
Seismograph 34
Settlements
bid rents 184, 186
factors affecting distribution 177
function 172
growth 171
hierarchy 173, 174
services 174
sites 170
sphere of influence 175, 177
see also Land-use/functional zones,
Socio-economic/ethnic areas
Shanty town 193, 194
Shock waves 36
Site
bridging point 170
defensive site 170
wet point 170
Social indicators 147
Socio-economic/ethnic areas 183,
190, 198
Soil 67, 71, 76
profile 72
South-west India, sustainable
development project 153
Spain, population density and
distribution 93
Spider diagrams 216
Stevenson's Screen 2
Sunrise industries, *see* Hi-tech
industries
Sustainability
development 152, 202
economic 128
ecotourism 85
need for solutions to globl
warming 29
urban environments 202
Synoptic Charts 12, 15

Taiga 67
Tanzania, assisting through aid 164
Technology 152
Temperature 2, 7, 8, 10, 11, 17, 18,
19
Thermometers 2
Threshold 174
Tourism 28, 64, 85
Town settlement, growth over time
171
Trade 156
Trade deficit 156
Trade surplus 156
Traffic control 210
Transitional zone 202
Transnational corporations (TNCs)
141, 143, 144

Underground features 60
Underpopulation 113
United Arab Emirates 104
Urban
development 185
environments, planning 202
fringe 172, 185, 187, 188, 203, 191
gentrification 203
land-use models 183
regeneration 202
residential zone 196, 202
sprawl 212

Urbanisation 179
causes of 181

Vegetation 67, 73, 75, 81
Vehicles, effect on environment
26
Volcanoes 31

Waste management 208

Weather 2
conditions 2
data 13
elements 2
forecasting 12
maps 226
maps 3, 8, 12
recording 2
recording instruments 2
station model 5
symbols 12
see also Climate
Weathering
biological 57
chemical 57
mechanical 57
Winds 3
direction 2
patterns 4
power 119
prevailing 18
speed 2, 4, 119
symbols 5
turbines 120
vane 2
World Summit for Sustainable
Development (WSSD) 29
World trade 156

Youth dependency 105